"Exploring Creation With Astronomy - Lessons 1-14 Lapbook Package

PLEASE NOTE: This product includes BOTH lapbooks for this book. One lapbook covers lessons 1-6, and the other covers lessons 7-14.

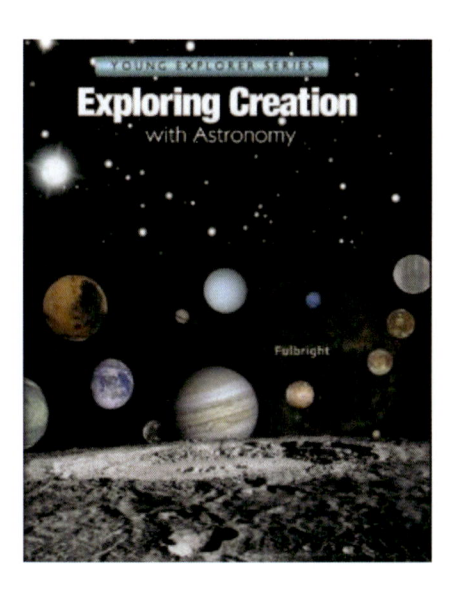

This lapbook has been specifically designed for use with the book, "Exploring Creation with Astronomy" by Jeannie Fulbright and Apologia Science.

Templates are printed with colors that best improve information retention according to scientific research.

Designed by
Cyndi Kinney
of Knowledge Box Central
with permission from Apologia Science
and Jeannie Fulbright.

Exploring Creation With Astronomy –
Lessons 1-14 Lapbook Package
Copyright © 2010 Knowledge Box Central
www.KnowledgeBoxCentral.com

ISBN #
Ebook:978-1-61625-046-1
CD:978-1-61625-047-8
Printed:978-1-61625-048-5

Publisher: Knowledge Box Central
Http://www.knowledgeboxcentral.com

All rights reserved. No part of this publication may be reproduced, stored in a retrieval system or transmitted in any form by any means, electronic, mechanical, photocopy, recording or otherwise, without the prior permission of the publisher, except as provided by USA copyright law.

The purchaser of the eBook or CD is licensed to copy this information for use with the immediate family members only. If you are interested in copying for a larger group, please contact the publisher.

Pre-printed or Pre-Assembled formats are not to be copied and are consumable. They are designed for one student only.

All information and graphics within this product are originals or have been used with permission from its owners, and credit has been given when appropriate. These include, but are not limited to the following: www.iclipart.com, and Art Explosion Clipart.

> This book is dedicated to my amazing family. Thank you to my wonderful husband, Scott, who ate a lot of leftovers, listened to a lot of whining (from me!), and sent lots of positive energy my way. Thank you to my daughter, Shelby, who truly inspired me through her love for learning. Thank you to my parents, Judy and Billy Trout, who taught me to trust in my abilities and to never give up.

L-A1

"Exploring Creation With Astronomy" - Lessons 1-6 Lapbook

PLEASE NOTE: This is the first of 2 lapbooks for this book. This particular lapbook ONLY covers Lessons 1 through 6, while another lapbook covers Lessons 7-14. You will need BOTH lapbooks in order to complete the entire book in lapbook format.

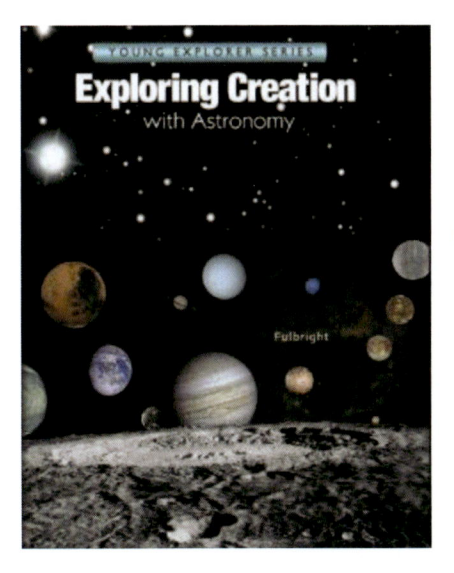

This lapbook has been specifically designed for use with the book, "Exploring Creation with Astronomy" by Jeannie Fulbright and Apologia Science.

Templates are printed with colors that best improve information retention according to scientific research.

Designed by
Cyndi Kinney
of Knowledge Box Central
with permission from Apologia Science
and Jeannie Fulbright.

Exploring Creation With Astronomy – Lessons 1-6 Lapbook
Copyright © 2010 Knowledge Box Central
www.KnowledgeBoxCentral.com

ISBN #
Ebook: 978-1-61625-046-1
CD: 978-1-61625-039-3
Printed: 978-1-61625-040-9
Assembled: 978-1-61625-041-6

Publisher: Knowledge Box Central
Http://www.knowledgeboxcentral.com

All rights reserved. No part of this publication may be reproduced, stored in a retrieval system or transmitted in any form by any means, electronic, mechanical, photocopy, recording or otherwise, without the prior permission of the publisher, except as provided by USA copyright law.

The purchaser of the eBook or CD is licensed to copy this information for use with the immediate family members only. If you are interested in copying for a larger group, please contact the publisher.

Pre-printed or Pre-Assembled formats are not to be copied and are consumable. They are designed for one student only.

All information and graphics within this product are originals or have been used with permission from its owners, and credit has been given when appropriate. These include, but are not limited to the following: www.iclipart.com, and Art Explosion Clipart.

> This book is dedicated to my amazing family. Thank you to my wonderful husband, Scott, who ate a lot of leftovers, listened to a lot of whining (from me!), and sent lots of positive energy my way. Thank you to my daughter, Shelby, who truly inspired me through her love for learning. Thank you to my parents, Judy and Billy Trout, who taught me to trust in my abilities and to never give up.

Table of Contents

How To Get Started …………………………………..………..5

Now What? ……………………………………………………..6

Color & Shapes ……………………………………...…………..7

Base Assembly & Layouts ……………………………………8-9

Pictures of Completed Lapbook …… ……………...10-11

Student Instruction Guide ……………………………..12-14

Lapbook Assembly Guide ……………………………...15-18

Teacher's Guide ……………………………………..…...19-21

Booklet Templates:

    Lesson 1:
        Booklet #1 …………………………………..22
        Booklet #2 …………………………………..23
        Booklet #3 …………………………………..24
        Booklet #4 …………………………………..25
        Booklet #5 …………………………………..26-28
        Booklet #6 (Vocabulary Booklet)……. ……..29

    Lesson 2:
        Booklet #1 …………………………………..30
        Booklet #2 …………………………...……..31, 33
        Booklet #3 …………………………………..32-33
        Booklet #4 …………………………………..24
        Booklet #5 …………………………………..34
        Vocabulary Booklet………..……… …….29

Lesson 3:
  Booklet #1 ……………………………………..35
  Booklet #2 ……………………………………..36
  Booklet #3 ……………………………………..34
  Vocabulary Booklet ………….…………….29

Lesson 4:
  Booklet #1 ……………………………….....37-38
  Booklet #2 ……………………………………..39
  Booklet #3 ……………………………………..40
  Booklet #4 ……………………………………..36
  Vocabulary Booklet ………….…………….29

Lesson 5:
  Booklet #1 …………………………….....41-44
  Booklet #2 …………………………………..45
  Booklet #3 …………………………………..46
  Booklet #4 …………………………………..47
  Booklet #5 (Vocabulary Booklet)………..…50

Lesson 6:
  Booklet #1 …………………………………..51-57
  Booklet #2 …………………………………….48
  Booklet #3 ……………………………………..58
  Booklet #4 ……………………………….59-60
  Booklet #5 ……………………………………..61
  Booklet #6 ……………………………………..62
  Vocabulary Booklet …………………….…..50

How do I get started?

First, you will want to gather your supplies.

*** Assembly:

    *Folders: We use colored file folders, which can be found at Walmart, Sam's, Office Depot, Costco, etc. You will need between 1 and 4 file folders, depending on which product you have purchased. You may use manila folders if you prefer, but we have found that children respond better with the brightly colored folders. Don't worry about the tabs....they aren't important. Within this product, you will be given easy, step-by-step instructions for how to fold and assemble these folders. *If you prefer, you can purchase the assembled lapbook bases from our website.*

    *Glue: For the folder assembly, we use hot glue. For booklet assembly, we use glue sticks and sometimes hot glue, depending on the specific booklet. We have found that bottle glue stays wet for too long, so it's not a great choice for lapbooking. For gluing the folders together, we suggest using hot glue, but ONLY with adult supervision. These things get SUPER hot, and can cause SEVERE burns within seconds.

    *Other Supplies: Of course, you will need scissors. Many booklets require additional supplies. Some of these include metal brad fasteners, paper clips, ribbon, yarn, staples, hole puncher, etc. You may want to add decorations of your own, including stickers, buttons, coloring pages, cut-out clipart, etc. Sometimes, we even use scrapbooking supplies. The most important thing is to use your imagination! Make it your own!!

Continue ON........ ⟶

Ok. I've gathered the supplies. Now how do I use this product?

Inside, you will find several sections. They are as follows:

1. **Lapbook Assembly Guide:** This section gives instructions and diagrams will tell the student exactly how to assemble the lapbook base and where to glue each booklet into the base. Depending on the student's age, he or she may need assistance with this process, especially if you choose to allow the student to use hot glue.

2. **Student Instruction Guide:** This section is written directly to the student, in language that he or she can understand. However, depending on the age of the child, there may be some parent/teacher assistance needed. This section will also tell the student exactly what should be written inside each booklet as he or she comes to it during the study, as well telling the student which folder each booklet will be glued into.

3. **Lapbook Assembly Guide:** This section is written directly to the student also, in language that he or she can understand. However, as with the previous section, depending on the age of the child, there may be some parent/teacher assistance needed. This section will also tell the student how to cut, fold, and assemble each booklet.

4. **Teacher's Guide**: This section is a great resource for the parent/teacher. In this section, you will find the page number where each answer may be found in the book. You will also find suggestions of extra activities that you may want to use with your student.

5. **Booklet Templates:** This section includes ALL of the templates for the booklets. These have been printed on colors that will help to improve retention of the information presented, according to scientific research on color psychology.

Colors & Shapes – Why Do They Matter?

After MUCH research and studies, science has shown that colors and shapes have psychological values. These influence the emotions and memories of each one of us. In our products, we have used specific colors and shapes in ways that will improve information retention and allow for a much more mentally interactive time of study. Some pages may have a notation at the bottom, where a specific color is suggested for your printing paper. This color suggestion is designed to improve information retention. However, if you do not have that specific color of paper, just print on whatever color you have. For the most benefit, follow the color suggestions, and watch your child's memory and enthusiasm truly soar!

Be creative!

Make it your own!

If you would like to send pictures of your completed lapbook, please do!

We would love to display your lapbooks on our website and/or in our newsletter.

Just send your pictures, first initial & last name, and age to us at: cyndi@knowledgeboxcentral.com

Exploring Creation With
Astronomy: Lessons 1-6
Base Assembly & Layouts

You will need 4 folders of any color. Take each one and fold both sides toward the original middle fold and make firm creases on these folds (Figure 1). Then glue (and staple if needed) the backs of the small flaps together (Figure 2).

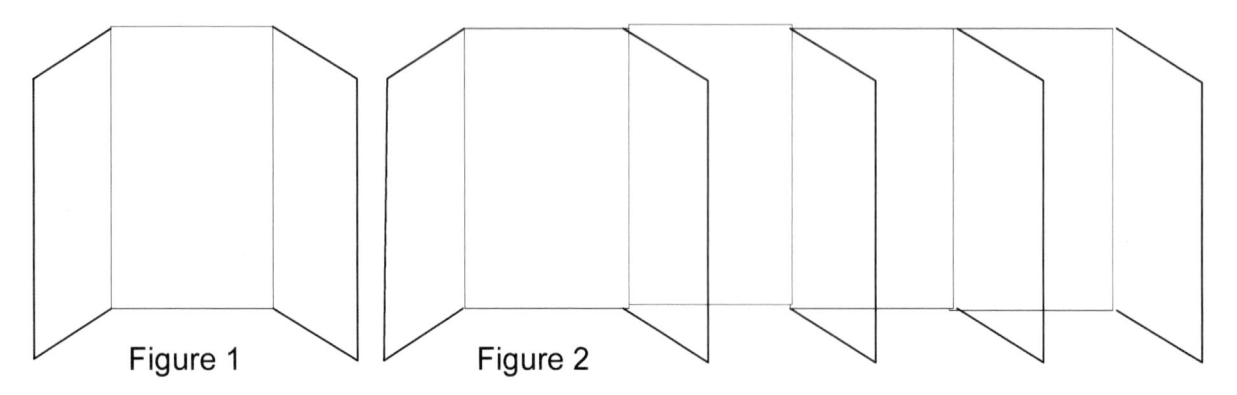

Figure 1

Figure 2

This is the "Layout" for your lapbook. The shapes are not exact on the layout, but you will get the idea of where each booklet should go inside your lapbook.

Inside of 1st Folder:

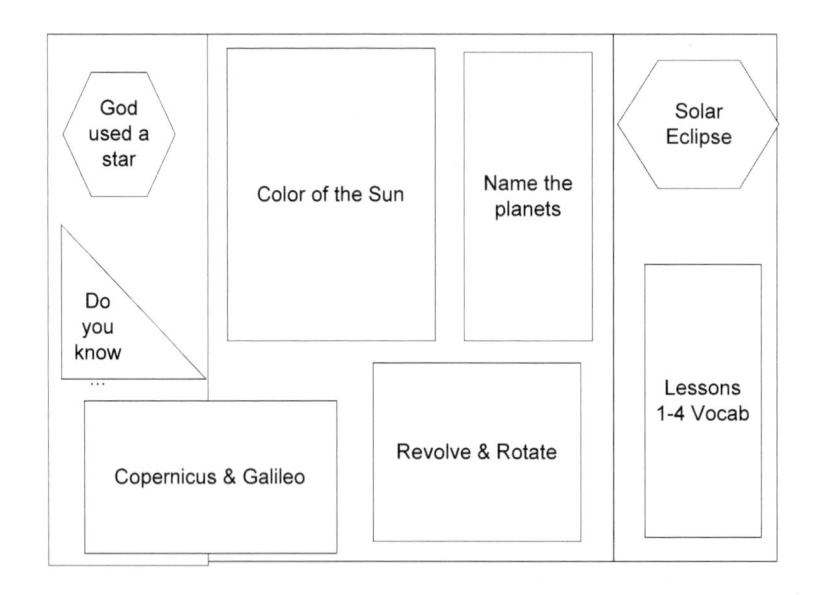

- God used a star
- Color of the Sun
- Name the planets
- Solar Eclipse
- Do you know…
- Copernicus & Galileo
- Revolve & Rotate
- Lessons 1-4 Vocab

Continue ON…….. ➡

Inside of 2nd Folder:

- Why is Venus hotter...
- Find Mercury
- Mercury's Craters
- Mercury Facts
- How many men..
- Fact
- Fact
- Phases of Venus
- Find Venus
- What would happen
- Venus
- Twins?

Inside of 3rd Folder:

- Lunar Eclipse
- Earth's Layers
- Phases of Our Moon
- Neil Armstrong
- Lessons 5-6 Vocab

Inside of 4th Folder:

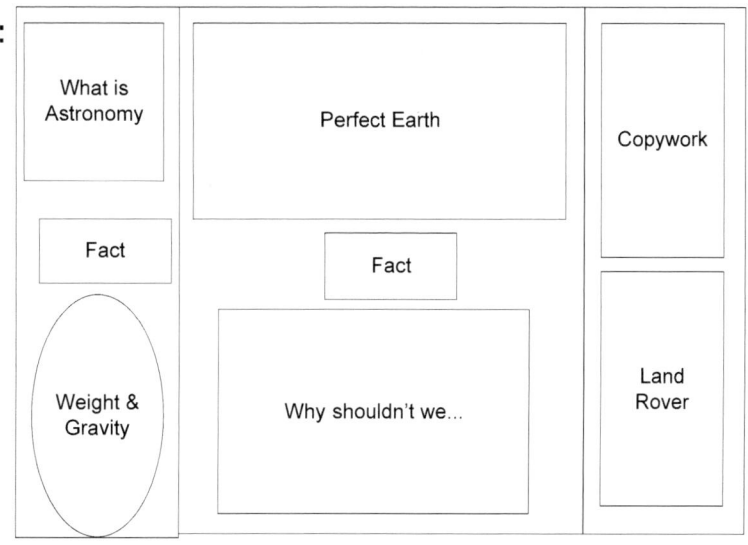

- What is Astronomy
- Fact
- Weight & Gravity
- Perfect Earth
- Fact
- Why shouldn't we...
- Copywork
- Land Rover

Below are pictures of a completed lapbook!!! This should help in figuring out how to assemble the booklets and then how to put it all together!

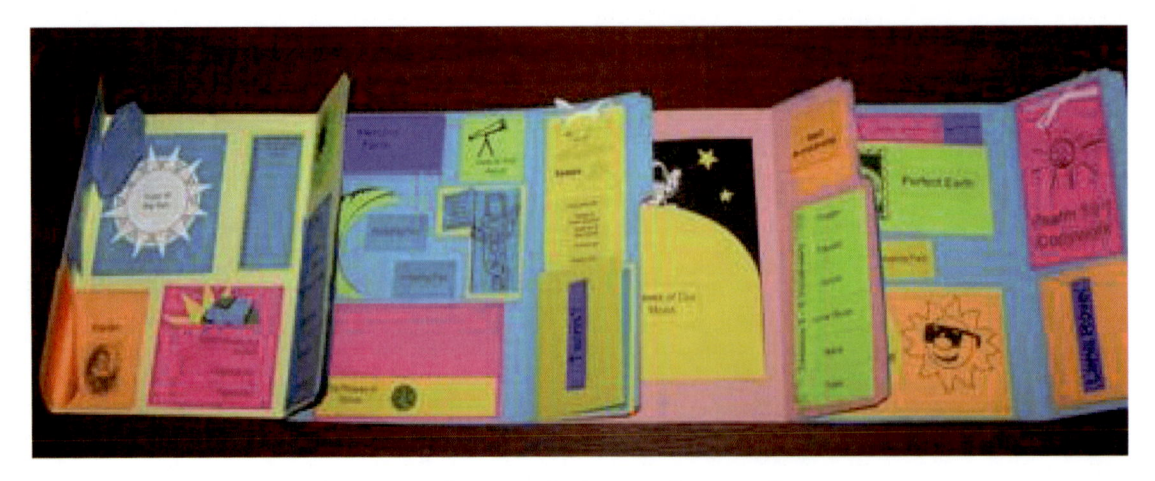

Completed Lapbook

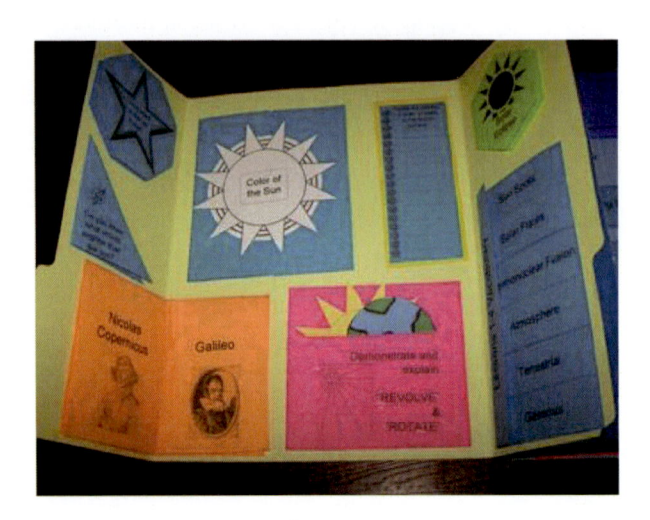

1st Folder

Continue ON........ ➡

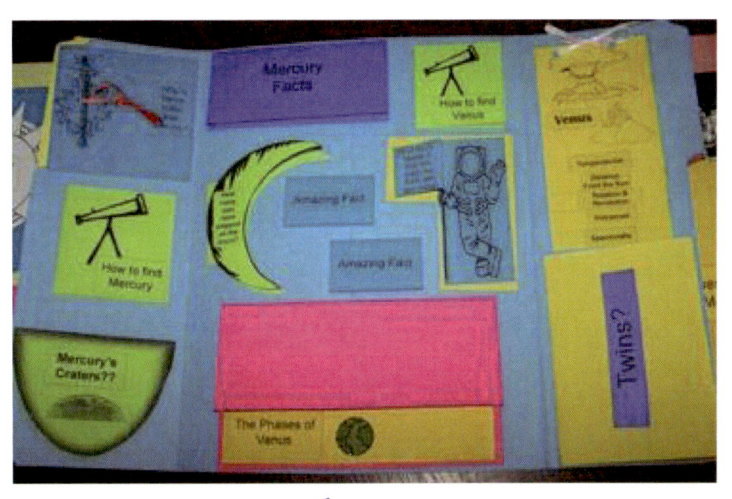

2nd Folder

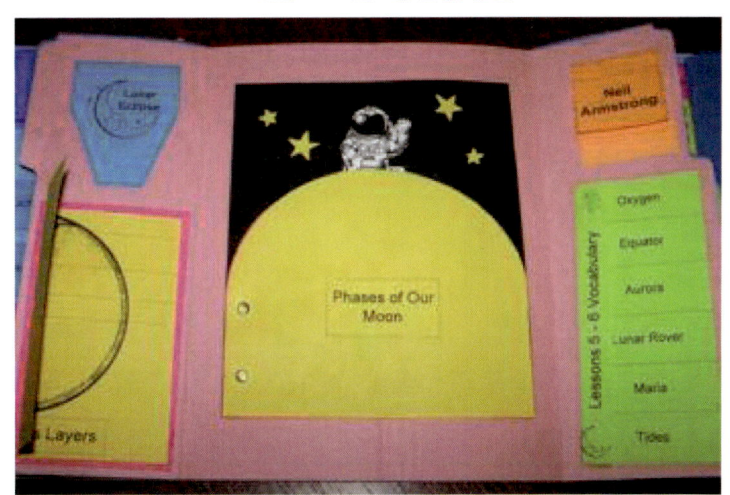

3rd Folder

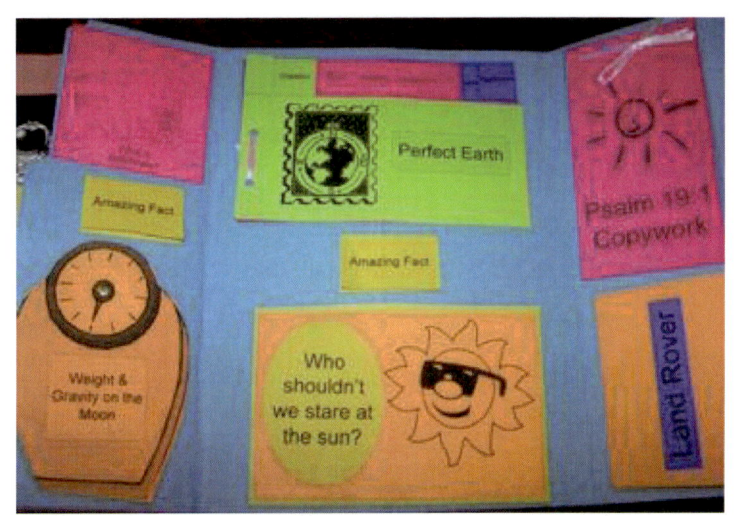

4th Folder

Exploring Creation with Astronomy
Lessons 1 - 6
Lapbook Student Instruction Guide

Lesson 1:
- Booklet 1 – What is Astronomy Booklet: This is a big word. Do you know what it means? Write the definition here, and color the picture on the outside of the booklet. (Folder 4)
- Booklet 2 – Star as a Sign Booklet: Do you remember the story about the birth of Jesus? God used a star as a special sign that night. Write about it here. (Folder 1)
- Booklet 3 – Name the Planets Booklet: Do you know the names of all of the planets? How about the order of them, starting with the closest to the sun? Try to write them all here, in order. (Folder 1)
- Booklet 4 – Copernicus & Galileo Booklet: These men had amazing ideas about our solar system many years ago. Tell about each one and what they believed. (Folder 1)
- Booklet 5 – Lessons 1-4 Vocabulary Booklet: This booklet contains words that you are learning. During lessons 1, 2, 3, and 4, you will write definitions inside this booklet. (Folder 1)
- Booklet 6 – Copywork Booklet: When you look at the sky and think about its awesome qualities, it is obvious that God is truly Almighty. This verse from Psalms will help you to remember. Use your best penmanship to copy the words into your booklet. (Folder 4)

Lesson 2:
- Booklet 1 – Why Should We Not Stare... Booklet: Did you know that your eyes can be damaged by looking directly at the sun? Why is this? Explain in this booklet. (Folder 4)
- Booklet 2 – Revolve & Rotate Booklet: Do you know the difference between "revolve" and "rotate?" They are very similar...yet very different and important in astronomy. Use the cut-outs of the sun and earth to demonstrate and explain the differences between these two words. (Folder 1)
- Booklet 3 – Color of the Sun Booklet: What color is the sun? The REAL answer may surprise you! Write the answer here, and color the picture of the sun on the front of the booklet. (Folder 1)
- Booklet 4 – What Shines Brighter... Booklet: Even though the sun shines amazingly bright...there is something that shines even brighter. Do you know what it is? (Folder 1)
- Booklet 5 – Solar Eclipse Booklet: Have you ever looked up toward the sun during the day....and it wasn't there? Explain how this can happen. (Folder 1)

- Remember: Lessons 1-4 Vocabulary Booklet: This booklet contains words that you are learning. During lessons 1, 2, 3, and 4, you will write definitions inside this booklet. (Folder 1)

Continue ON........ ➡

Lesson 3:
o Booklet 1 – Mercury Facts Booklet: Have you ever seen Mercury? Use this booklet to tell about all of the things your have learned about this planet……..from temperatures to rotation/revolution to visiting spacecraft and more! (Folder 2)
o Booklet 2 – Mercury's Craters Booklet: Have you ever seen a crater? Craters are found on many planets...but not Mercury. Explain why this is true. (Folder 2)
o Booklet 3 – How to Find Mercury Booklet: If you wanted to look at Mercury, where would you look? Explain here. (Folder 2)
o REMEMBER – Lessons 1-4 Vocabulary Booklet: This booklet contains words that you are learning. During lessons 1, 2, 3, and 4, you will write definitions inside this booklet. (Folder 1)

Lesson 4:
o Booklet 1 – Venus Facts Booklet: Venus is a very hot planet! One reason is because of the volcanoes. Color the volcanoes on the front of this booklet, and then tell what you know about other features of this planet. (Folder 2)
o Booklet 2 – Twins? Booklet: There are many similarities between Earth and Venus. Some have even called them "twins." However, there are many differences as well. Use this venn diagram to show these similarities and differences. (Folder 2)
o Booklet 3 – Phases of Venus Booklet: Have you ever looked at Venus and noticed that it doesn't always look the same? Why does it change? Use this booklet to tell about the different phases. Draw them on this chart. (Folder 2)
o Booklet 4 – How to Find Venus Booklet: Have you ever seen Venus? If you were going to tell a friend how to find Venus in the sky, what would you tell him? (Folder 2)
o REMEMBER – Lessons 1-4 Vocabulary Booklet: This booklet contains words that you are learning. During lessons 1, 2, 3, and 4, you will write definitions inside this booklet. (Folder 1)

Lesson 5:
o Booklet 1 – Perfect Earth Booklet: Did you know that you live on a "perfect" planet? Well...there are a lot of things that don't seem "perfect," but the way the planet is made is very perfect. The greatest designer in the world planned it all out and put it together so that we could survive. In this booklet, you will explain all of the wonderful characteristics that make the Earth "perfect." (Folder 4)
o Booklet 2 – What Would Happen If…. Booklet: Since we have already said that God made the Earth to be "perfect," let's look at what would happen if He hadn't. Imagine what life would be like if He had made the Earth with less mass. What would happen to the gravity, and how different would your life be? (Folder 2)
o Booklet 3 – Earth's Layers Booklet: Our Earth has many layers. Can you name and identify them? (Folder 3)
o Booklet 4 – Why is Venus Hotter Booklet – Do you know why Venus is hotter than Mercury? Explain inside this booklet (Folder 2)
o Booklet 5 – Lessons 5-6 Vocabulary Booklet: Use this booklet to explain the definitions of some of the new words that you are learning. (Folder 3)

Lesson 6:
o Booklet 1 – Phases of the Moon Booklet: Have you noticed that the moon changes shape? There is a very important reason that this happens, and it was all in God's master plan. In this booklet, you will write the names of the phases, color in the phases, and track the moon for a month. Check it every day! (Folder 3)
o Booklet 2 – Lunar Eclipse Booklet: Did you know that the moon can seem to disappear? How can this happen? Explain it here. (Folder 3)
o Booklet 3 – How Many Men? Booklet: Can you believe that humans have been to the moon? How many have gone? (Folder 2)

Continue ON…….. ➡

o   Booklet 4 -   Weight on the Moon Booklet: Did you know that your weight on the Earth and on the Moon are very different? What would your weight be on the moon?? (Folder 4)

o   Booklet 5 - Neil Armstrong Booklet: Have you ever heard of this man? What was his great accomplishment in the world of Astronomy? (Folder 3)

o   Booklet 6 - Lunar Rover Booklet: This is a very odd-looking mode of transportation, wouldn't you say? Narrate how it was used to your teacher, and color the picture. (Folder 4)

o   REMEMBER - Lessons 5-6 Vocabulary Booklet: Use this booklet to explain the definitions of some of the new words that you are learning. (Folder 3)

Exploring Creation With Astronomy
Lessons 1 - 6
Lapbook Assembly Guide

Inside of 1st Folder:

1. "God used a star" Booklet: Cut out along the outer black lines, and fold in the center, so that the title is on the front.

2. "Do you know.." Booklet: Cut out along the outer black lines, and fold in the center, so that the title is on the front.

3. Copernicus & Galileo Booklet: Cut out along the outer black line edges. Then, fold along the center vertical line, so that the words are on the outside. Now, cut along the vertical line that runs between the 2 names on the front. This line only goes to the fold. You will place this booklet in a fold. See picture:

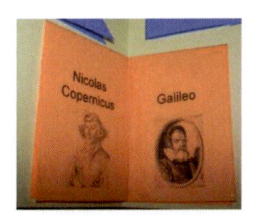

4. Color of the Sun Booklet: Cut out along the outer black line edges of the booklet and also of the sun shape (with the title on it). The sun shape will be on cardstock. Fold along each of the 2 dotted vertical lines, so that the edges meet each other in the center. Glue the sun on ONLY ONE SIDE of the booklet – where the curved line is. See picture:

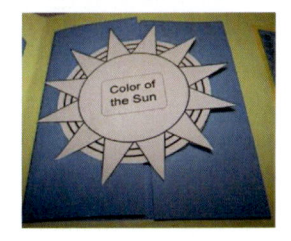

5. Name the Planets Booklet: Cut out along the outer black line edges. Now, glue to a piece of paper that is slightly larger and of a different color, creating a small border all the way around.

6. Revolve & Rotate Booklet: Cut out along the outer black line edges. Then, fold along the center vertical line, creating a "pocket." Glue only the outer edge and bottom edge of this pocket, so that there is an opening at the top, where you will place the sun and earth shapes. On cardstock, you will have the sun and earth shapes. Cut them out, and place them into the pocket. See picture:

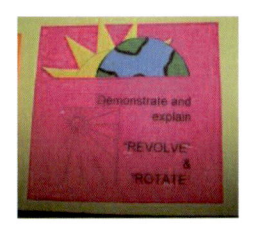

7. Solar Eclipse Booklet: Cut out along the outer black line edges. Fold along the center line, so that the title is on the front.

Continue ON…….. ➡

8. Lessons 1-4 Vocabulary Booklet: Cut out along the outer black line edges of the booklet. Fold along the vertical line nearest the center, so that the words are on the outside. Then, fold along the other vertical line, so that the title "flap" will fold slightly over the vocabulary words. Then fold along the horizontal lines that are between the words, creating little "flaps." See picture:

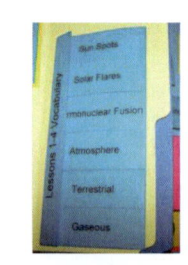

Inside of 2nd Folder:
1. "Why is Venus Hotter…" Booklet: Cut out along the outer black line edges. Fold along the 2 vertical lines, so that the rounded edges almost touch in the front (title on front too). Punch a hole through each of the small circles in the center, and loosely tie with a ribbon. See picture:

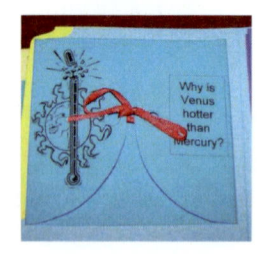

2. "Find Mercury" Booklet: Cut out along the outer black line edges. Fold along the center line so that the title is on the front.

3. Mercury's Craters Booklet: Cut out along the outer black line edges, keeping the half-circles connected. Then fold along the center line (where the sides meet) so that the title is on the front.

4. Mercury Facts Booklet: Cut out along the outer black line edges of the booklet and its title box. Accordion-fold the booklet, so that there is a blank on top and bottom. All of the words should be on the inside… .until you glue on the title. Glue the title on the outside top of the booklet. See picture:

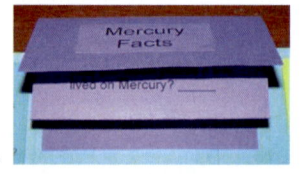

5. Find Venus Booklet: Cut out along the outer black line edges of the booklet. Fold along the vertical center line so that the title is on the front.

6 "How many men…" booklet: Cut out along the outer edges of the booklet, keeping the "moons" connected. Then, fold in the center between them, with the title ending up on the front.

7. Amazing Facts Booklets (2): Cut out along the outer black line edges. Fold along the center lines, so that the titles are on the fronts.

8. "What would happen if…" Booklet: Cut out along the outer edges of the astronaut and his flag. Also cut out the question text box. Fold the flag along the center line so that the flag opens from the front. Glue the question to the front of the flag booklet. Now glue the entire thing to a piece of paper of a different color. Trim around the edges, leaving a small border around the edges. See picture:

Continue ON……..

9. Phases of Venus Booklet: Cut out along the outer black line edges of the page. Glue to a piece of paper of a different color. Trim around the edges so that there is a small border all the way around. Fold horizontally so that only the title and globe show. See picture:

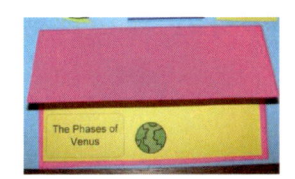

10. Venus Booklet: Cut out each page along the outer black line edges. Note that each page is a little longer than the one before. Stack the pages so that the title is on top and each page gets progressively longer toward the back. Punch 2 holes in the top, and secure with a ribbon or yarn. See picture:

11. Twins Booklet: Cut out along the outer black line edges of the venn diagram page and the title text box. Glue the venn diagram to a piece of paper of a different color. Trim around the edges, so that there is a small border around the edges. Fold in the center, vertically. Glue the title to the outside. You will only glue one side of the booklet to the folder. See picture:

Inside of 3rd Folder:
1. Lunar Eclipse Booklet: Cut along the outer black line edges of the booklet. Fold along the center horizontal line, so that the title is on the front.
2. Earth's Layers Booklet: Cut along the outer black line edges. Now, mount to another piece of paper of a different color. Cut around the edges, leaving a small border.
3. Phases of Our Moon Booklet: This booklet is tricky. Pay attention to the instructions, and look at the pictures. Cut out each page of the booklet along the outer black line edges. Cut out the stars and astronaut from the cardstock. There is also a page that is a large rectangle, and you should use that as a template for cutting out the same shape on a black or dark blue piece of paper. Stack the pages with the rounded edges together, so that the title is on top. Now stack them on the black/blue page at the very bottom edge. Punch 2 holes in the left edge of the pages, and secure with metal brad fasteners. Now glue the astronaut to the black/blue page, just above the edge of the other pages. Also glue the stars around the astronaut. See picture above:

4. Neil Armstrong Booklet: Cut out along the outer black line edges of the booklet and the picture. Fold along each of the lines, so that each overlaps the other, and you end up with the title on the top. Glue the picture in the center of the booklet when it is opened all the way up. See picture:

5. Lesson 5-6 Vocabulary Booklet: Cut out and assemble as you did with the Lesson 1-4 Vocabulary booklet.

Continue ON……..

Inside of 4th Folder:

1. Astronomy Booklet: Cut out along the outer black line edges. Fold along the middle line, so that the title is on the front.

2. Amazing Fact Booklets (2): Cut out along the outer black line edges. Fold along the center lines so that the titles are on the fronts.

3. Weight & Gravity Booklet: Cut out along the outer black line edges of each page. Stack together so that the title is on the front. Punch a hole through the center of the scale's face (through all pages). Then secure with fastener. See picture:

4. Perfect Earth Booklet: Cut out along the black line edges of each page. Notice that each page has a "tab" at the top, and each "tab" is longer than the other. Stack them on top of each other, so that the title is on top, and each page's "tab" gets progressively longer until the back page is a full page. Punch 2 holes on the far left side, and secure them with a ribbon or yarn. See picture:

5. "Why shouldn't we stare…" Booklet: Cut out along the outer black line edge of the page with the sun and also the booklet. Glue the booklet with the words on it to the left of the sun. Now, glue the entire thing to another piece of paper of a different color. Trim around the edges, leaving a small border. See picture:

6. Copywork Booklet: Cut out along the outer black line edges of each page. Stack them so that the title is on top and the Psalm is second….with lined pages are next. Punch 2 holes in the top of the stack, and secure with ribbon or yarn.

7. Land Rover Booklet: Cut out along the outer edges of the picture and the title box. Secure to a slightly larger piece of paper of a different color, creating a small border around the edges. Fold in half, vertically. Secure the title to the outside. Glue only one side to the folder. The booklet assembles, folds, and is glued in just like the "Twins" booklet.

Exploring Creation With Astronomy
Lessons 1 – 6 Lapbook
Teacher's Guide

Here, you'll find information to supplement your study. Jeannie Fulbright's book is so wonderfully filled with wisdom and creation confirmation. All of the information needed to complete all of the booklets can be found on the pages of her book. Below, I will tell you which pages hold specific "answers." Also, you'll find many other sites listed, where you may want to go for extra information, coloring pages, games, crafts, and ideas to extend your study.

I have been questioned as to *why I merely give you the page numbers for the answers instead of the answers themselves*. If I were to give you ONLY the answers, then there would be no need for you to have Jeannie's awesome book...right? Also, this will require the parent to actually read the book as well, which was Jeannie's intent from the beginning. So, I hope that you understand my decision to not "just give the answers." It really is a calculated plan on mine and Jeannie's part.

Lesson 1 :
- Booklet 1 – What is Astronomy Booklet: Answer on page 2
- Booklet 2 – Star as a Sign Booklet: Answer on page 4
- Booklet 3 – Name the Planets Booklet: Answer on pages 5-6
- Booklet 4 – Copernicus & Galileo Booklet: Answers on page 7
- Booklet 5 – Lessons 1-4 Vocabulary Booklet: Answers throughout all 4 Lessons
- Booklet 6 – Copywork Booklet

Additional Resources for Lesson 1:
* Information about Copernicus and his theories: http://csep10.phys.utk.edu/astr161/lect/retrograde/copernican.html
* Short biography about Galileo: http://www.lucidcafe.com/library/96feb/galileo.html
* Enchanted Learning's Zoom site: http://www.enchantedlearning.com/subjects/astronomy/planets/
* Astronomy Word Search: http://www.puzzles.ca/wordsearch/astronomy.html

Lesson 2:
- Booklet 1 – Why Should We Not Stare... Booklet : Answers on pages 13-14
- Booklet 2 – Revolve & Rotate Booklet : Answers on page 14
- Booklet 3 – Color of the Sun Booklet : Answers on pages 20-22
- Booklet 5 – What Shines Brighter... Booklet : Answer on page 23
- Booklet 6 – Solar Eclipse Booklet : Answers found on pages 23-25
- REMEMBER – Lessons 1-4 Vocabulary Booklet: Answers throughout all 4 Lessons

Continue ON........ ➡

Additional Resources for Lesson 2:
* Solar Eclipse Activity/Model: http://solar-center.stanford.edu/eclipse/model.html
* Interactive Solar Eclipse Activity: http://teacher.scholastic.com/activities/science/moon_interactives.htm
* The Sun – picture and facts: http://www.nineplanets.org/sol.html
* Really cool sun foldable game: http://www.sec.noaa.gov/info/Origami.pdf
* Great site with lots of information about the sun: http://www.eyeonthesky.org/ourstarsun.html

Lesson 3:
o   Booklet 1 – Mercury Facts Booklet: Answers found on pages 30-35
o   Booklet 2 – Mercury's Craters Booklet: Answers found on pages 32-35
o   Booklet 3 – How to Find Mercury Booklet: Answer found on page 36
o   REMEMBER – Lessons 1-4 Vocabulary Booklet: Answers found throughout all 4 lessons

Additional Resources for Lesson 3:
* Information about Mercury: http://www.adlerplanetarium.org/cyberspace/planets/mercury/

Lesson 4:
o   Booklet 1 – Venus Facts Booklet: Answers found on pages 40-46
o   Booklet 2 – Twins? Booklet: Answer found on page 44
o   Booklet 3 – Phases of Venus Booklet: Answers found on pages 46-47
o   Booklet 4 – How to Find Venus Booklet: Answer found on page 47
o   REMEMBER – Lessons 1-4 Vocabulary Booklet: Answers found throughout all 4 Lessons

Additional Resources for Lesson 4:
* Interactive Phases of Venus: http://galileoandeinstein.physics.virginia.edu/more_stuff/flashlets/PhasesofVenus.htm
* Lots of info about bumble bees: http://www.everythingabout.net/articles/biology/animals/arthropods/insects/bees/bumble_bee/

Lesson 5:
o   Booklet 1 - Perfect Earth Booklet: Answers found on pages 52-62
o   Booklet 2 – What Would Happen If…. Booklet: Answer found on pages 53-54
o   Booklet 3 – Earth's Layers Booklet: Answer found on page 61
o   Booklet 4 – Why is Venus Hotter Booklet: Answer found on page
o   Booklet 5 – Lessons 5-6 Vocabulary Booklet: Answers found throughout both Lessons

Additional Resources for Lesson 5:
* Earth's Layers (interactive): http://scign.jpl.nasa.gov/learn/plate1.htm

Continue ON…….. ➡

Lesson 6:
- Booklet 1 – Phases of the Moon Booklet: Answers found on pages 67-69
- Booklet 2 – Lunar Eclipse Booklet: Answers found on pages 69-70
- Booklet 3 – How Many Men? Booklet: Answers found on pages 71-72
- Booklet 4 – Weight on the Moon Booklet: Answer found on page 72
- Booklet 5 – Neil Armstrong Booklet: Answer found on page 71-72
- Booklet 6 – Lunar Rover Booklet: Answer found on pages 71-72
- REMEMBER – Lessons 5-6 Vocabulary Booklet: Answers found throughout both Lessons

Additional Resources for Lesson 6:
* Earth's Layers (interactive): http://scign.jpl.nasa.gov/learn/plate1.htm
* Space Travel word search: http://sunniebunniezz.com/puzzles/spacedws.htm

Additional Resources for ALL LESSONS:
* Lots of great information: http://coolcosmos.ipac.caltech.edu/cosmic_classroom/ask_astronomer/faq/
* More great information, sponsored by NASA: http://image.gsfc.nasa.gov/poetry/ask/askmag.html
* Lots of great information and games/activities for kids: http://starchild.gsfc.nasa.gov/docs/StarChild/StarChild.html
* Really fun games from NASA: http://www.nasa.gov/audience/forkids/kidsclub/flash/index.html
* More fun games: http://science.hq.nasa.gov/index.html
* This site has TONS of fun games and information: http://www.kidsastronomy.com/
* Another great website for lots of fun: http://www.dustbunny.com/afk/
* And another fun site: http://www.astronomy.com/asy/default.aspx?c=a&id=1091

Lesson 1: Booklet 4

Nicolas Copernicus

Galileo

Lesson 1:
Booklet 5
(continued on next page)

Psalm 19:1 Copywork

Lesson 1: Booklet 5 (continued on next page)

Psalm 19:1

The heavens are telling of the glory of God; and their expanse is declaring the work of His hands."

Lesson 1: Booklet 5 (last page)

Lesson 1: Booklet 5 (also used in lessons 2, 3, & 4)

Sun Spots
Solar Flares
Thermonuclear Fusion
Atmosphere
Terrestrial
Gaseous

Lessons 1-4 Vocabulary

**Lesson 2:
Booklet 1**

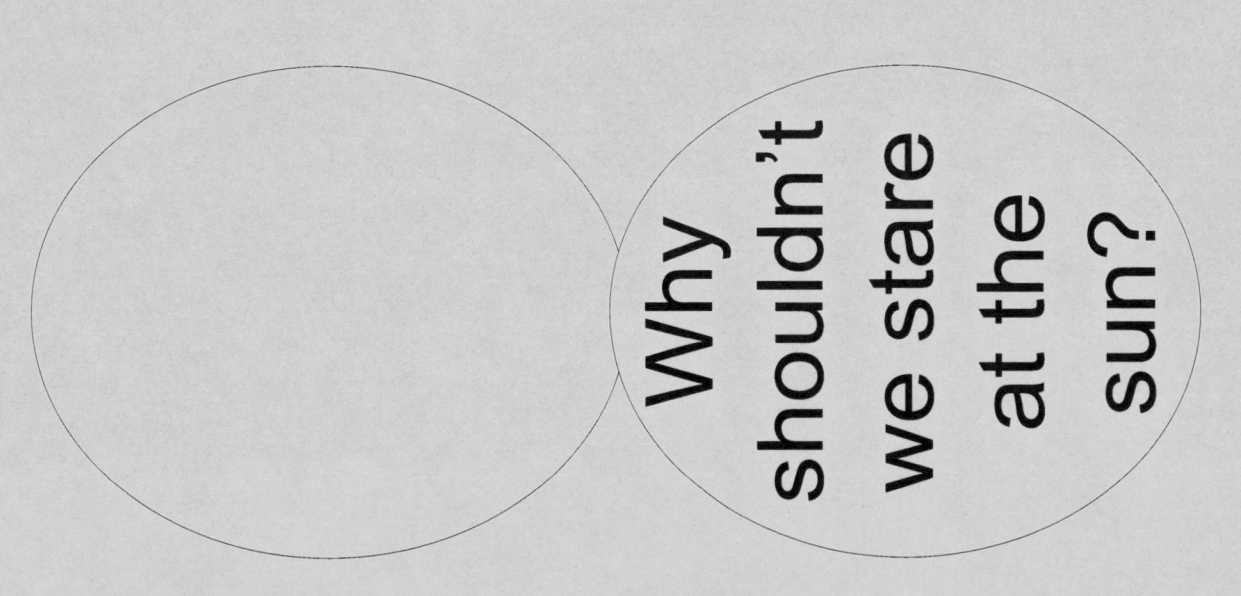

Why shouldn't we stare at the sun?

Lesson 2:
Booklet 2
(Part 1)

Demonstrate and explain

"REVOLVE"
&
"ROTATE"

Lesson 2: Booklet 3 (Part 1)

"Color of the Sun" booklet – sun with title is on cardstock page with "Revolve & Rotate" sun and earth.

Color of the Sun

Lesson 2: Booklet 3 (Part 2)

Lesson 2: Booklet 2 (Parts 2-4)

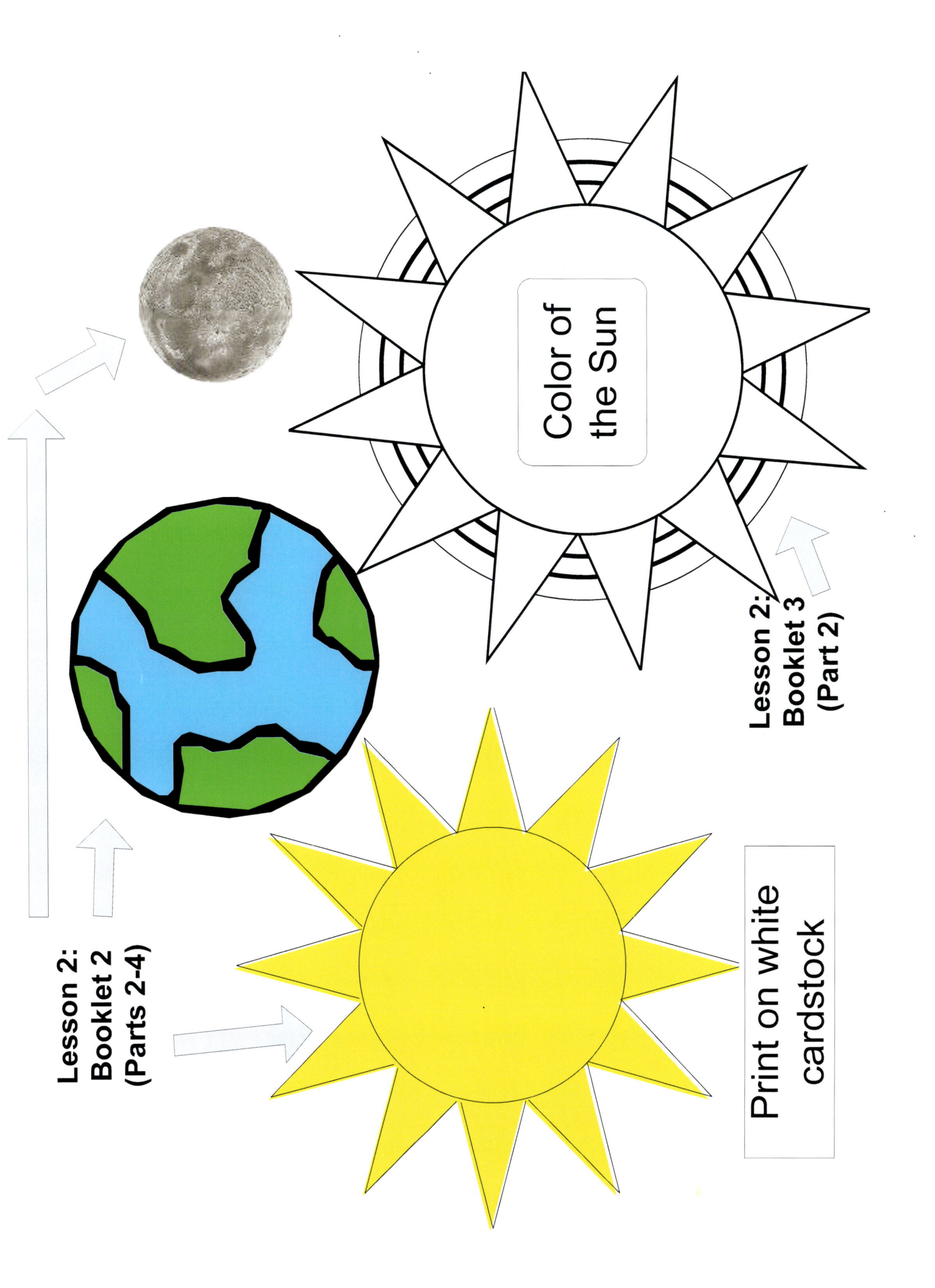

Print on white cardstock

Solar
Eclipse

Lesson 2: Booklet 5

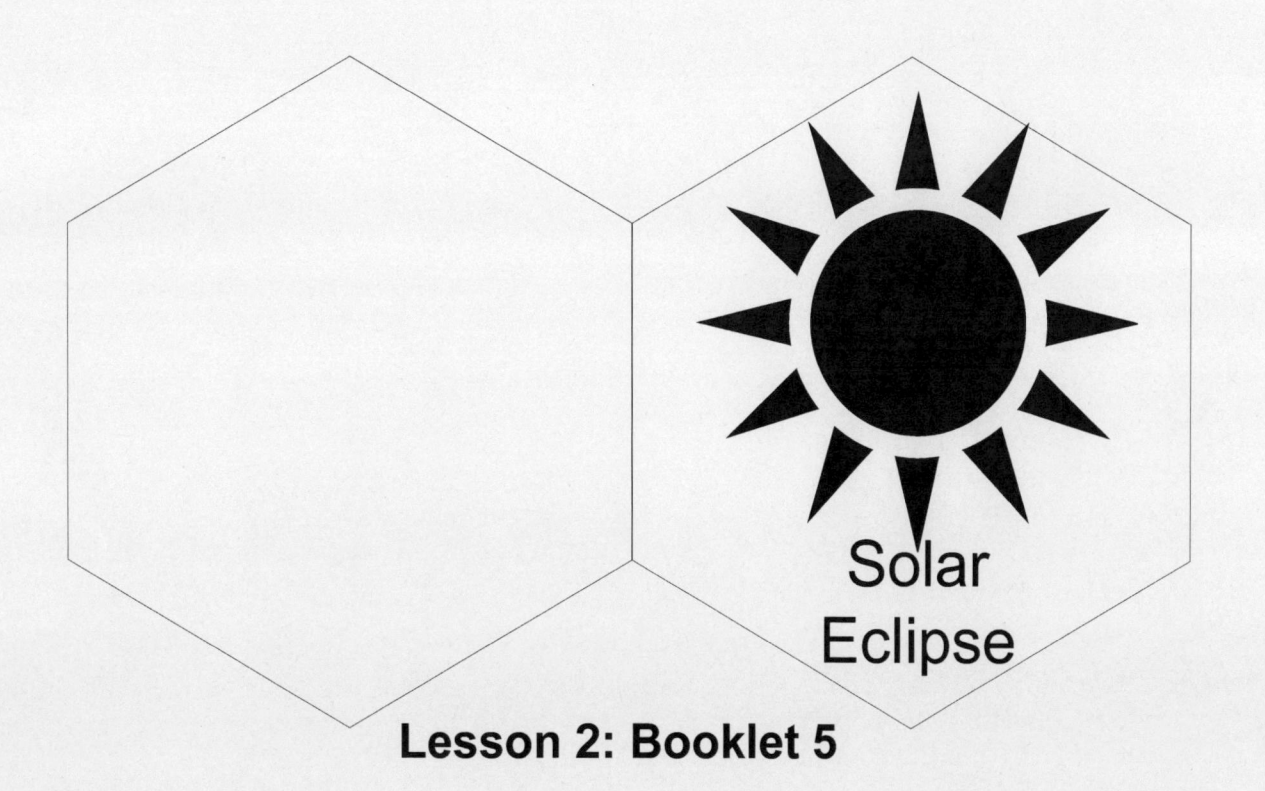

How to find
Mercury

Lesson 3: Booklet 3

Lesson 3: Booklet 1

Temperature on Mercury during the day: _____
Temperature on Mercury at night: _____
How many earth days is one day on Mercury? _____

How far from the sun is Mercury? _____
How old would you be if you lived on Mercury? _____

Mercury Facts

What is the name of the spacecraft that has been to Mercury?
_____

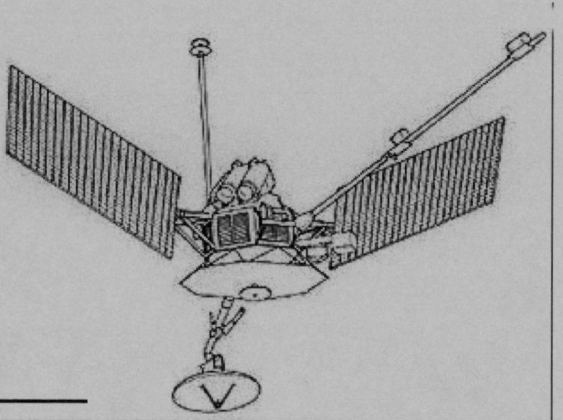

More information about Mercury:

Mercury's Craters??

**Lesson 3:
Booklet 2**

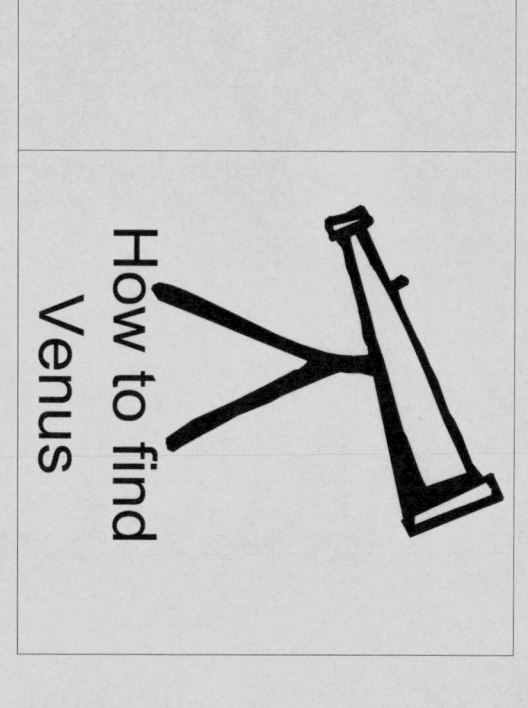

How to find Venus

**Lesson 4:
Booklet 4**

Venus

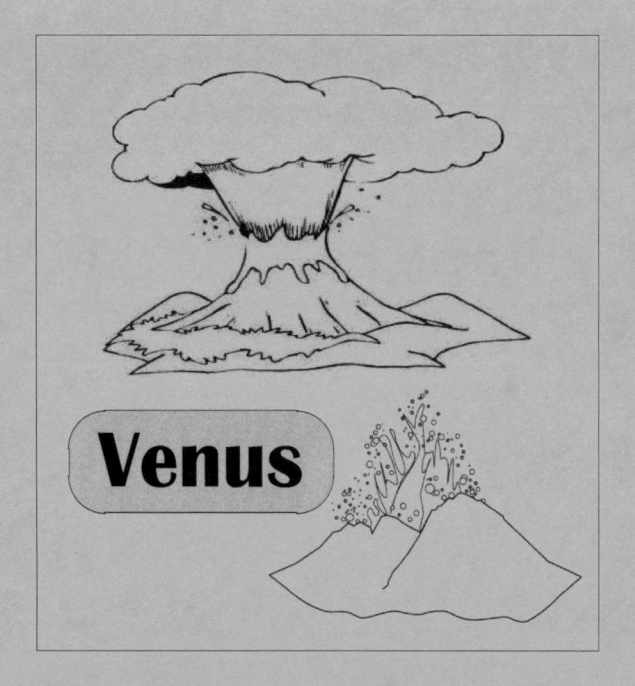

Temperatures

Distance From the Sun

Rotation & Revolution

**Lesson 4: Booklet 1
(continued on next page)**

Volcanoes

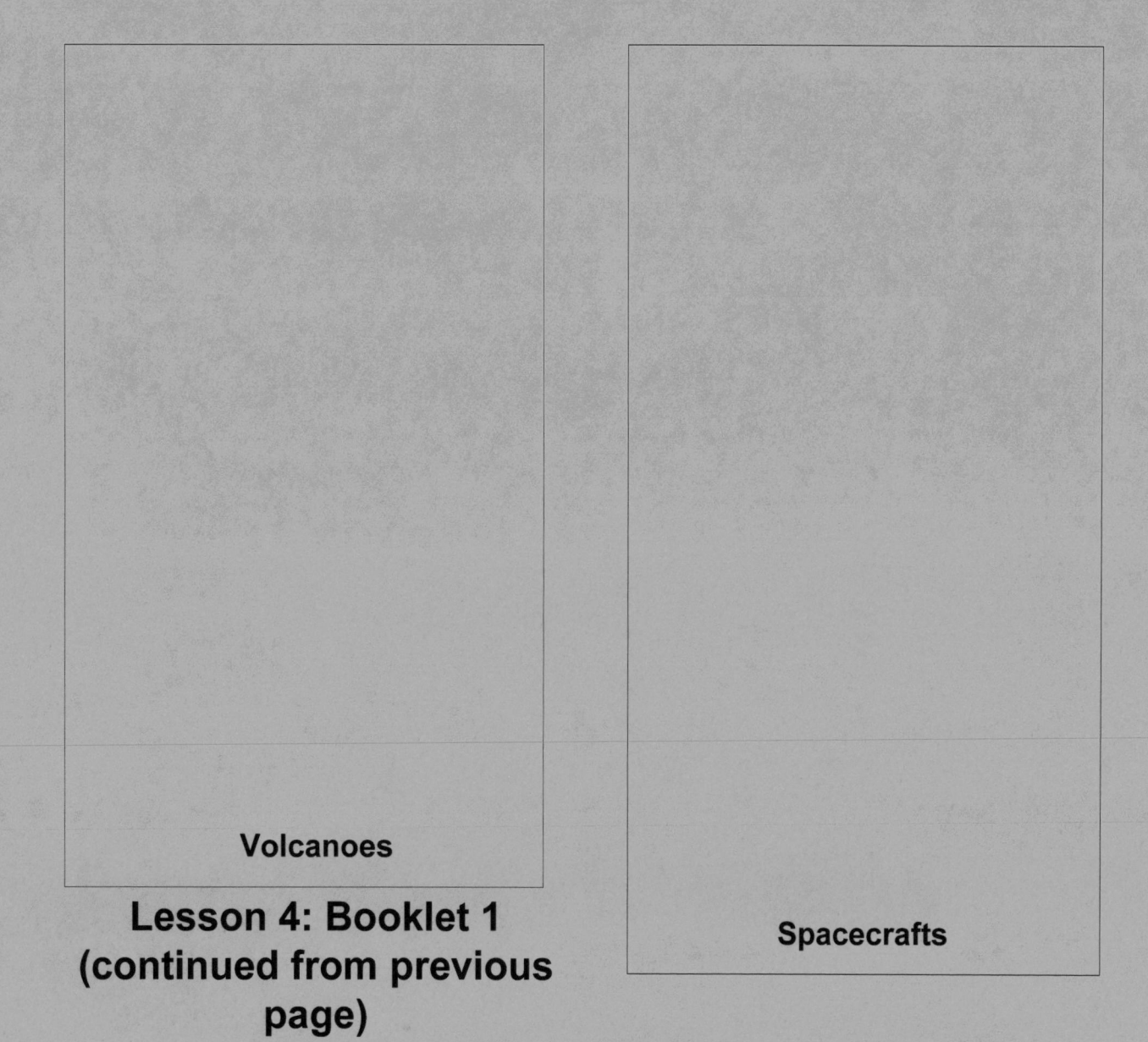

Lesson 4: Booklet 1 (continued from previous page)

Spacecrafts

Twins?

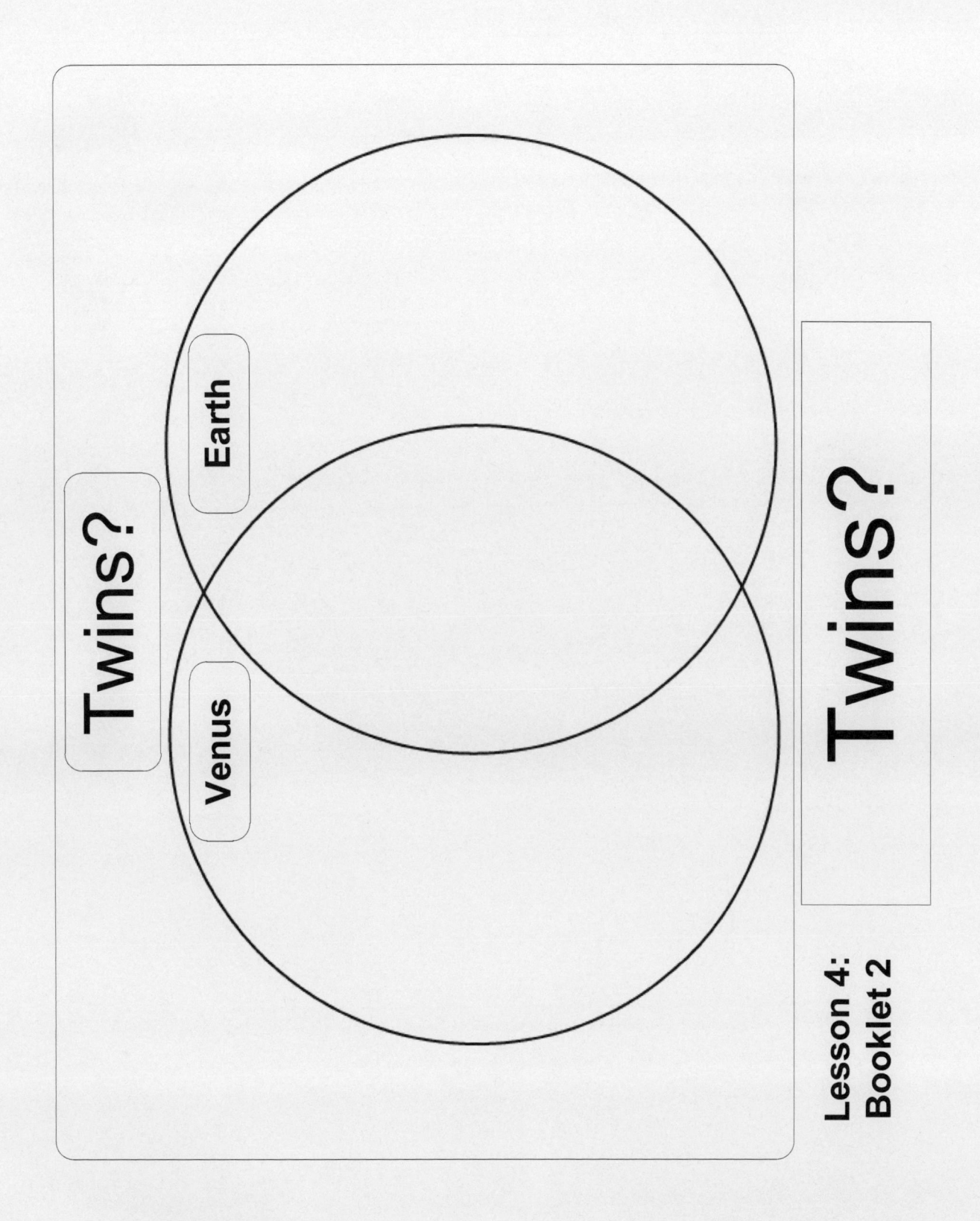

Earth

Venus

Twins?

**Lesson 4:
Booklet 2**

The Phases of Venus

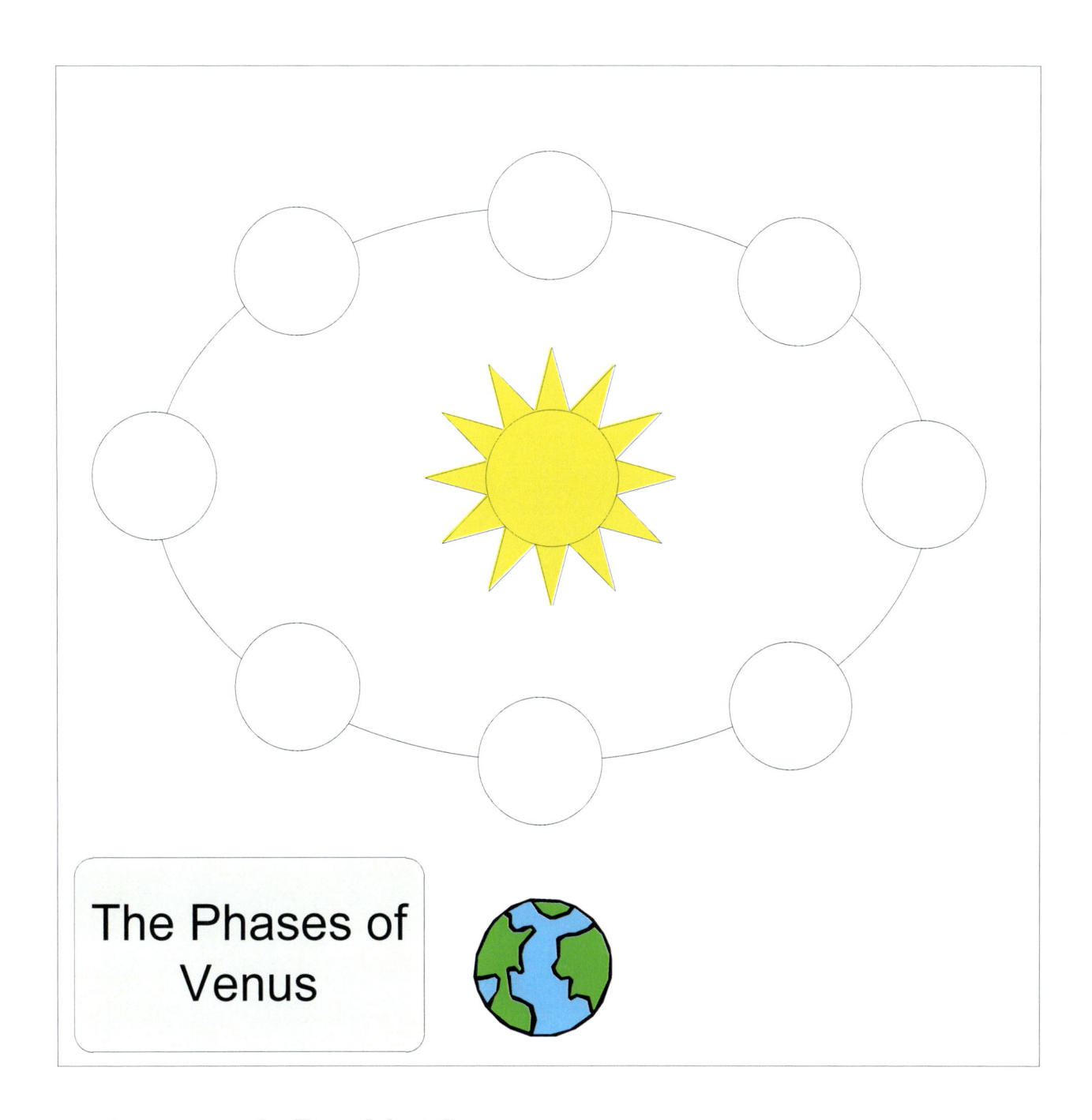

Lesson 4: Booklet 3

Suggested Paper Color: White

Lesson 5: Booklet 1 (continued on next 3 pages)

Perfect Earth

Distance

Mass

Lesson 5: Booklet 1

Rotation

Atmosphere

Lesson 5: Booklet 1

Tilt

Lesson 5: Booklet 1

Land

Magneto-
sphere

Lesson 5:
Booklet 2

What would happen if God had made the Earth with less mass?

**Lesson 5:
Booklet 3**

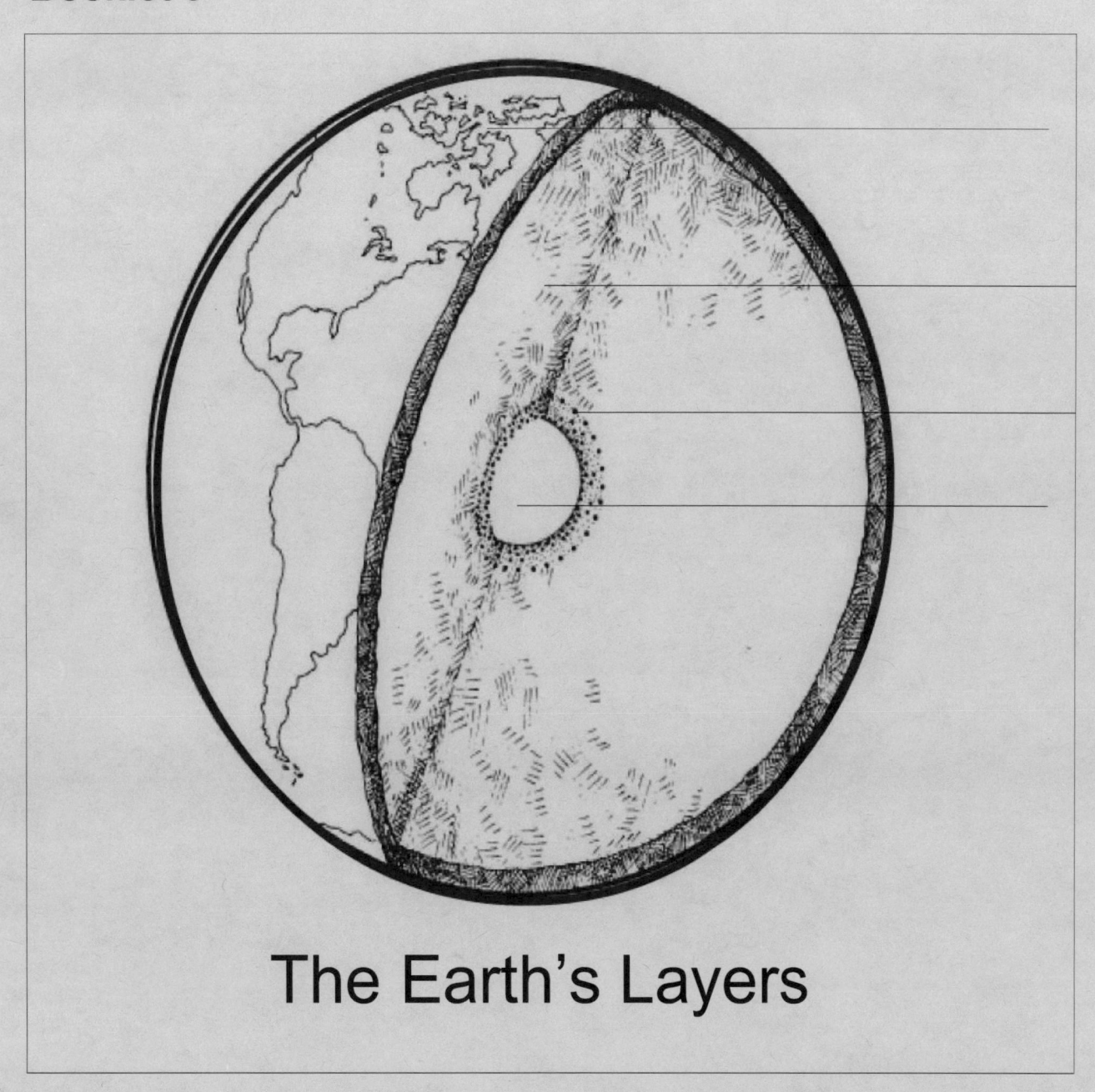

The Earth's Layers

Suggested Paper Color: Blue

**Lesson 5:
Booklet 4**

Why is Venus hotter than Mercury?

Suggested Paper Color: Red

**Lesson 6:
Booklet 2**

Lunar Eclipse

Amazing Fact	Amazing Fact
Amazing Fact	Amazing Fact

These are used in several lessons

| Oxygen |
| Equator |
| Aurora |
| Lunar Rover |
| Maria |
| Tides |

Lessons 5 - 6 Vocabulary

**Lesson 5:
Booklet 5**

**Lesson 6:
Booklet 1
(Part 1)**

Phases of
Our Moon

Suggested Paper Color: Yellow

Lesson 6: Booklet 1 (Part 2)

Label the Phases of the Moon

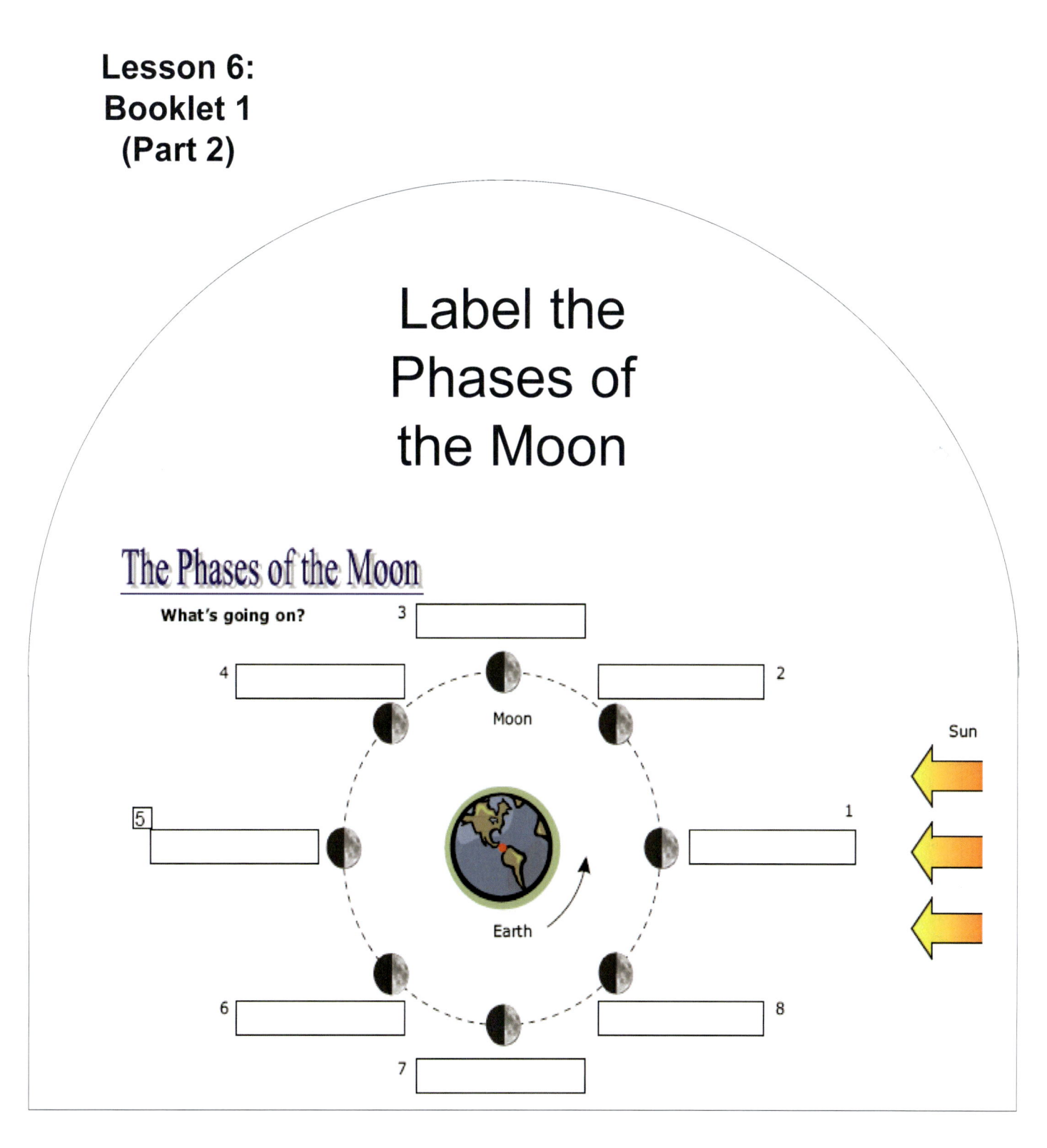

The Phases of the Moon

What's going on?

Sun

Moon

Earth

1.
2.
3.
4.
5.
6.
7.
8.

Suggested Paper Color: White

Lesson 6: Booklet 1 (Part 3)

Label/ Explain the Phases

Explanations:

Lesson 6: Booklet 1 (Part 4)

Track the moon for one month.

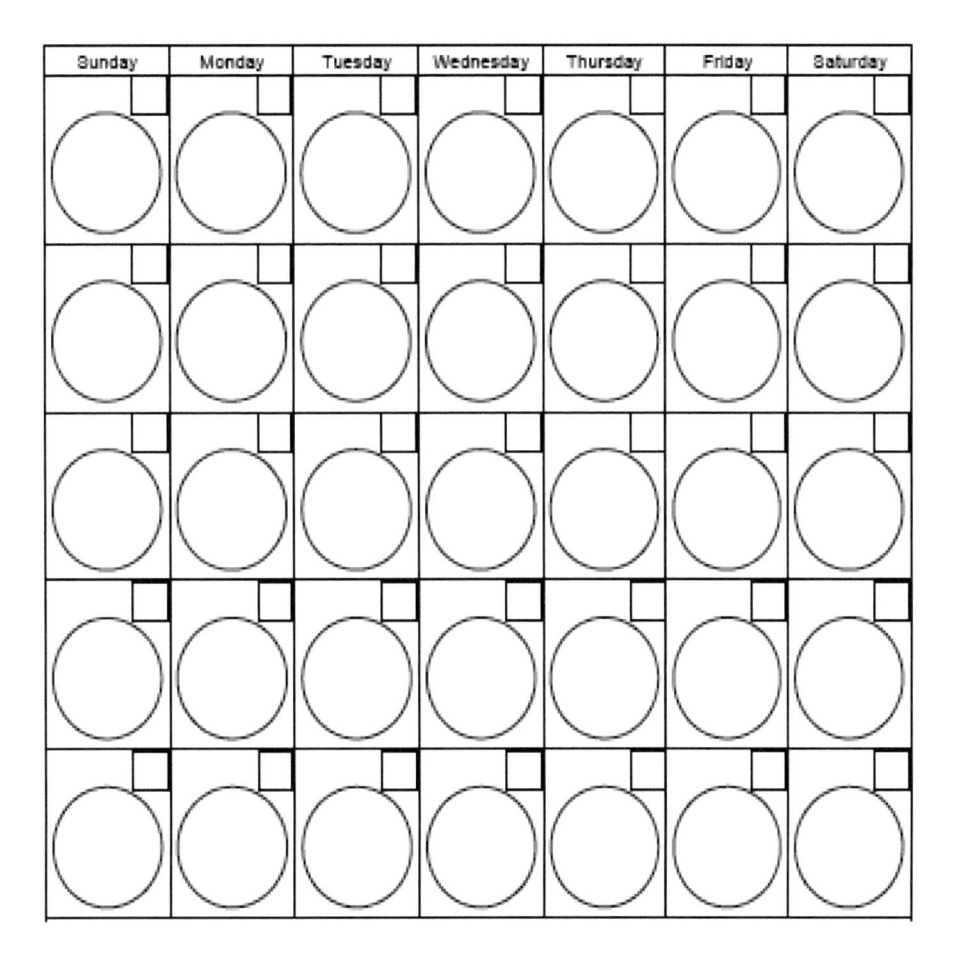

Sunday	Monday	Tuesday	Wednesday	Thursday	Friday	Saturday
○	○	○	○	○	○	○
○	○	○	○	○	○	○
○	○	○	○	○	○	○
○	○	○	○	○	○	○
○	○	○	○	○	○	○

Suggested Paper Color: White

Lesson 6: Booklet 1
(Part 5)

Print on white cardstock – cut out and use on Moon Booklet (see instructions)

Lesson 6: Booklet 1 (Part 6) | Back Page – print out this template, and use it to cut this shape from dark blue or black construction paper

Print on yellow – cut out and use on Moon Booklet (see instructions)

**Lesson 6:
Booklet 1
(Part 7)**

**Lesson 6:
Booklet 3**

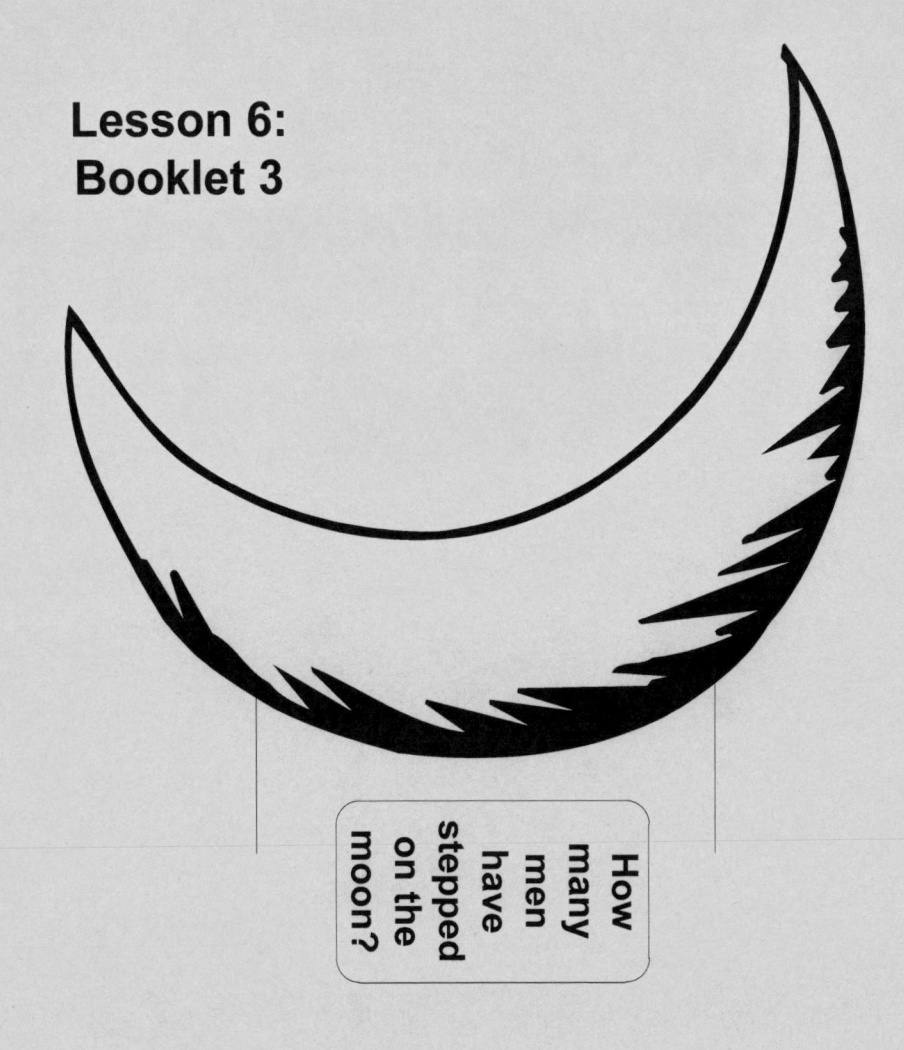

How many men have stepped on the moon?

Suggested Paper Color: Yellow

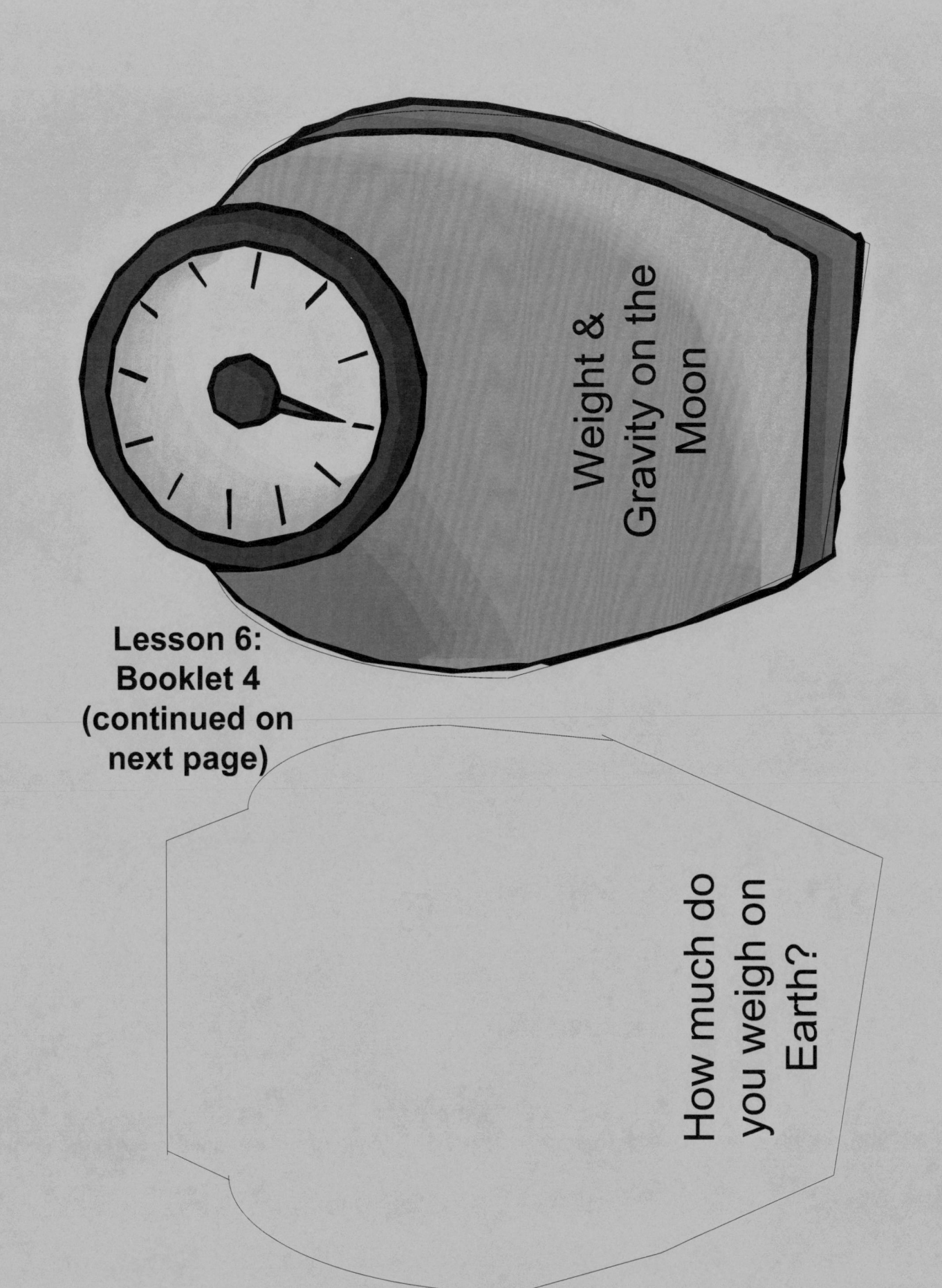

**Lesson 6:
Booklet 4
(continued on
next page)**

Weight & Gravity on the Moon

How much do you weigh on Earth?

How much would you weigh on the Moon?

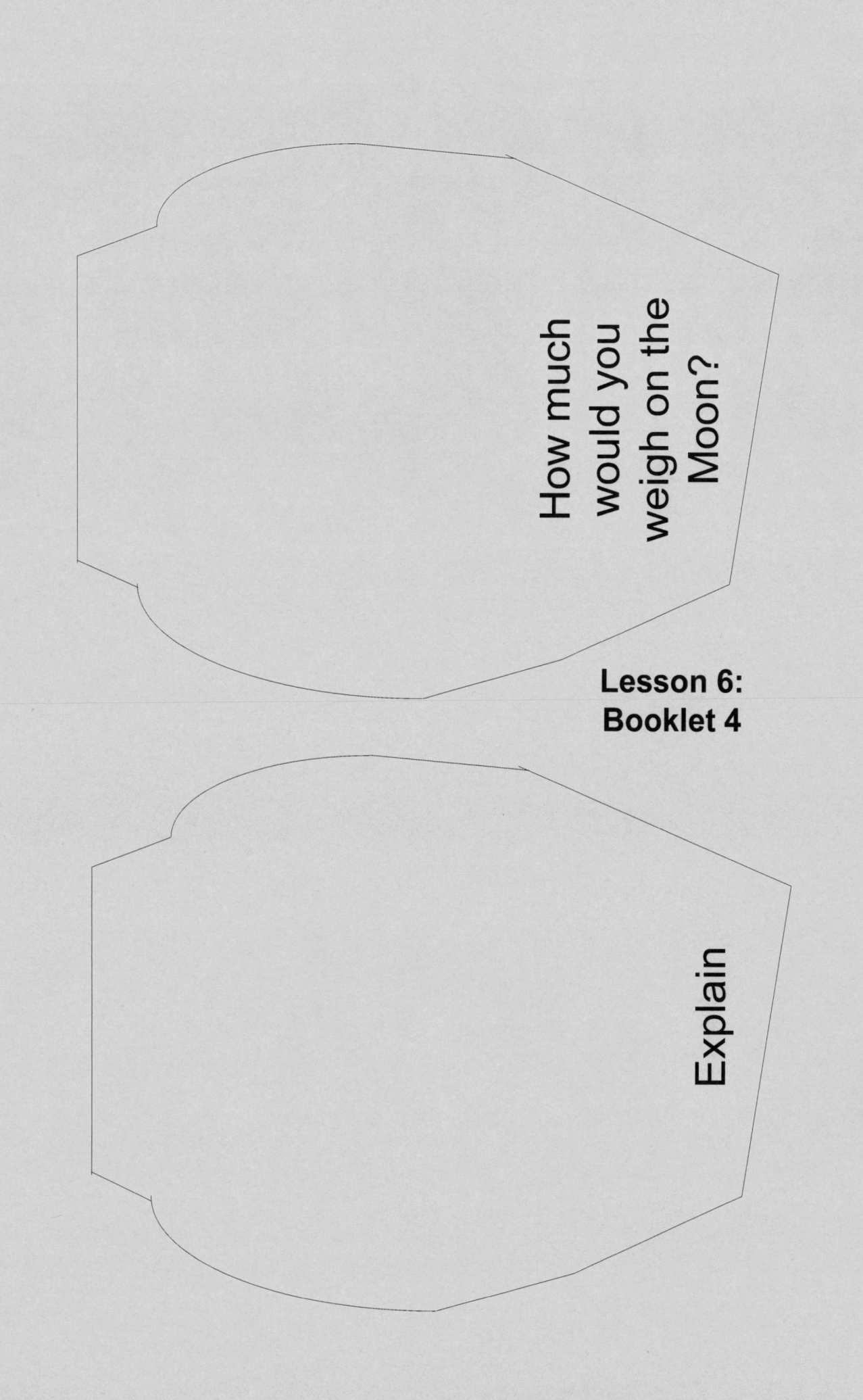

Lesson 6:
Booklet 4

Explain

Lesson 6:
Booklet 5

Name of spaceship

Date of 1st steps on the moon

Neil Armstrong

Lesson 6:
Booklet 6

LUNAR ROVER VEHICLE

Land Rover

L-A2

"Exploring Creation With Astronomy" - Lessons 7-14 Lapbook

PLEASE NOTE: This is the first of 2 lapbooks for this book. This particular lapbook ONLY covers Lessons 7-14, while another lapbook covers Lessons 1-6. You will need BOTH lapbooks in order to complete the entire book in lapbook format.

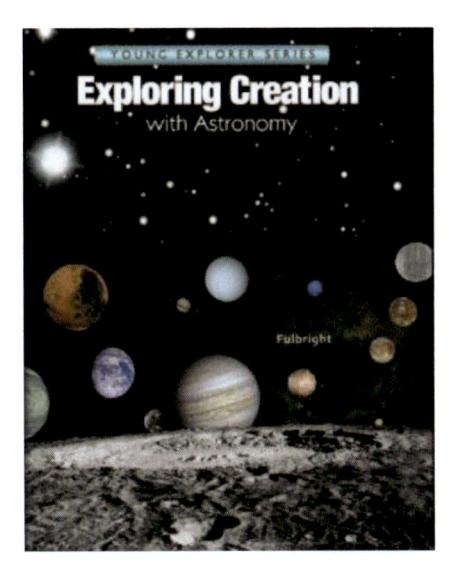

This lapbook has been specifically designed for use with the book, "Exploring Creation with Astronomy" by Jeannie Fulbright and Apologia Science.

Templates are printed with colors that best improve information retention according to scientific research.

Designed by
Cyndi Kinney
of Knowledge Box Central
with permission from Apologia Science
and Jeannie Fulbright.

Exploring Creation With Astronomy – Lessons 7-14 Lapbook
Copyright © 2010 Knowledge Box Central
www.KnowledgeBoxCentral.com

ISBN #
Ebook: 978-1-61625-042-3
CD: 978-1-61625-043-0
Printed: 978-1-61625-044-7
Assembled: 978-1-61625-045-4

Publisher: Knowledge Box Central
Http://www.knowledgeboxcentral.com

All rights reserved. No part of this publication may be reproduced, stored in a retrieval system or transmitted in any form by any means, electronic, mechanical, photocopy, recording or otherwise, without the prior permission of the publisher, except as provided by USA copyright law.

The purchaser of the eBook or CD is licensed to copy this information for use with the immediate family members only. If you are interested in copying for a larger group, please contact the publisher.

Pre-printed or Pre-Assembled formats are not to be copied and are consumable. They are designed for one student only.

All information and graphics within this product are originals or have been used with permission from its owners, and credit has been given when appropriate. These include, but are not limited to the following: www.iclipart.com, and Art Explosion Clipart.

> This book is dedicated to my amazing family. Thank you to my wonderful husband, Scott, who ate a lot of leftovers, listened to a lot of whining (from me!), and sent lots of positive energy my way. Thank you to my daughter, Shelby, who truly inspired me through her love for learning. Thank you to my parents, Judy and Billy Trout, who taught me to trust in my abilities and to never give up.

Table of Contents

How To Get Started …………………………..………4

Now What? ……………………………………….…..5

Color & Shapes …………………………...………...6

Base Assembly & Layouts ………………………… 7-8

Pictures of Completed Lapbook …… ……………....9-10

Student Instruction Guide ……………………….11-12

Lapbook Assembly Guide …………………….…...13-16

Teacher's Guide ………………………….…..…17-19

Booklet Templates…………………………..……...20-70

How do I get started?

First, you will want to gather your supplies.

*** Assembly:

    *Folders: We use colored file folders, which can be found at Walmart, Sam's, Office Depot, Costco, etc. You will need between 1 and 4 file folders, depending on which product you have purchased. You may use manila folders if you prefer, but we have found that children respond better with the brightly colored folders. Don't worry about the tabs….they aren't important. Within this product, you will be given easy, step-by-step instructions for how to fold and assemble these folders. ***If you prefer, you can purchase the assembled lapbook bases from our website.***

    *Glue: For the folder assembly, we use hot glue. For booklet assembly, we use glue sticks and sometimes hot glue, depending on the specific booklet. We have found that bottle glue stays wet for too long, so it's not a great choice for lapbooking. For gluing the folders together, we suggest using hot glue, but ONLY with adult supervision. These things get SUPER hot, and can cause SEVERE burns within seconds.

    *Other Supplies: Of course, you will need scissors. Many booklets require additional supplies. Some of these include metal brad fasteners, paper clips, ribbon, yarn, staples, hole puncher, etc. You may want to add decorations of your own, including stickers, buttons, coloring pages, cut-out clipart, etc. Sometimes, we even use scrapbooking supplies. The most important thing is to use your imagination! Make it your own!!

Continue ON…….. ⟶

Ok. I've gathered the supplies. Now how do I use this product?

Inside, you will find several sections. They are as follows:

1. **Lapbook Assembly Guide:** This section gives instructions and diagrams will tell the student exactly how to assemble the lapbook base and where to glue each booklet into the base. Depending on the student's age, he or she may need assistance with this process, especially if you choose to allow the student to use hot glue.

2. **Student Instruction Guide:** This section is written directly to the student, in language that he or she can understand. However, depending on the age of the child, there may be some parent/teacher assistance needed. This section will also tell the student exactly what should be written inside each booklet as he or she comes to it during the study, as well telling the student which folder each booklet will be glued into.

3. **Lapbook Assembly Guide:** This section is written directly to the student also, in language that he or she can understand. However, as with the previous section, depending on the age of the child, there may be some parent/teacher assistance needed. This section will also tell the student how to cut, fold, and assemble each booklet.

4. **Teacher's Guide**: This section is a great resource for the parent/teacher. In this section, you will find the page number where each answer may be found in the book. You will also find suggestions of extra activities that you may want to use with your student.

5. **Booklet Templates:** This section includes ALL of the templates for the booklets. These have been printed on colors that will help to improve retention of the information presented, according to scientific research on color psychology.

Colors & Shapes – Why Do They Matter?

After MUCH research and studies, science has shown that colors and shapes have psychological values. These influence the emotions and memories of each one of us. In our products, we have used specific colors and shapes in ways that will improve information retention and allow for a much more mentally interactive time of study. Some pages may have a notation at the bottom, where a specific color is suggested for your printing paper. This color suggestion is designed to improve information retention. However, if you do not have that specific color of paper, just print on whatever color you have. For the most benefit, follow the color suggestions, and watch your child's memory and enthusiasm truly soar!

BE CREATIVE!

Make it your own!

If you would like to send pictures of your completed lapbook, please do!

We would love to display your lapbooks on our website and/or in our newsletter.

**Just send your pictures, first initial & last name, and age to us at:
cyndi@knowledgeboxcentral.com**

Exploring Creation With
Astronomy: Lessons 7-14
Lapbook Assembly Guide

You will need 4 folders of any color. Take each one and fold both sides toward the original middle fold and make firm creases on these folds (Figure 1). Then glue (and staple if needed) the backs of the small flaps together (Figure 2).

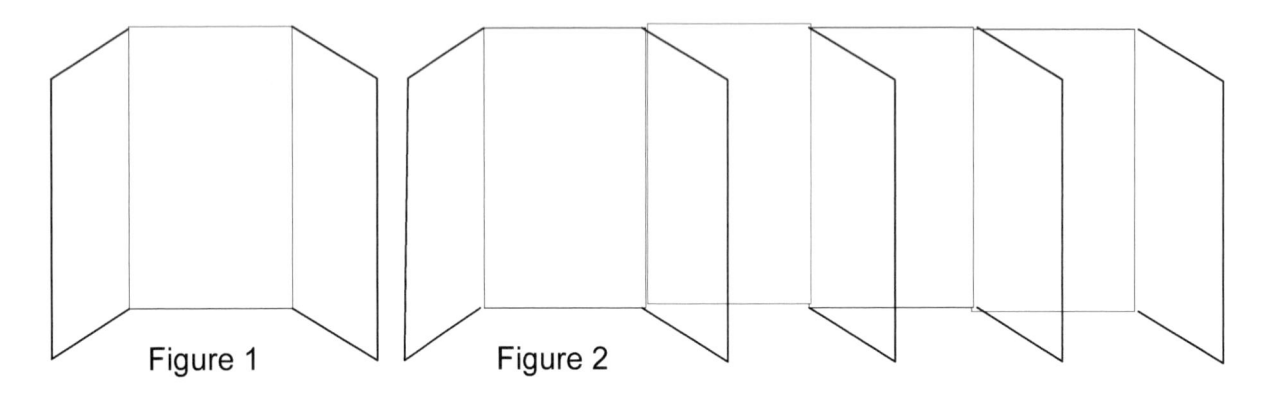

Figure 1

Figure 2

This is the "Layout" for your lapbook. The shapes are not exact on the layout, but you will get the idea of where each booklet should go inside your lapbook.

Inside of 1st Folder:

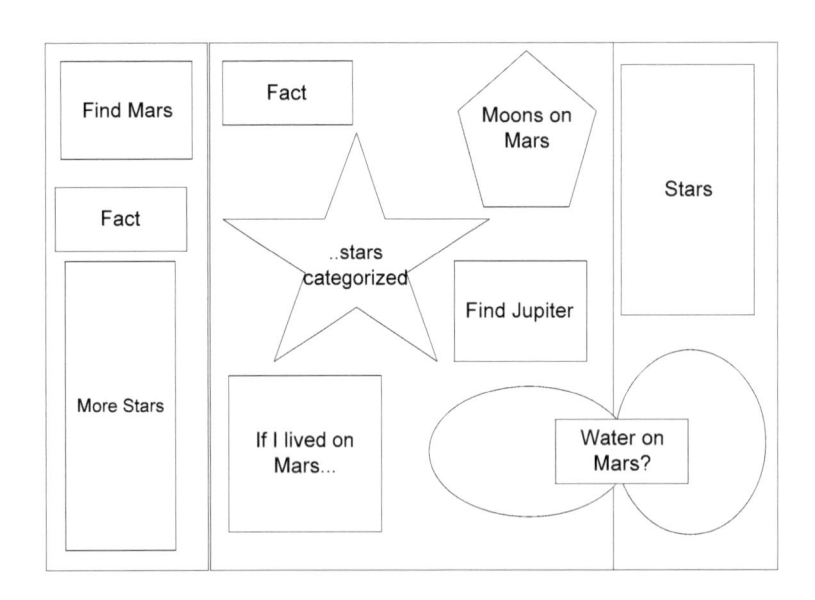

- Find Mars
- Fact
- Fact
- Moons on Mars
- Stars
- More Stars
- ..stars categorized
- Find Jupiter
- If I lived on Mars...
- Water on Mars?

Continue ON........ ➡

Inside of 2nd Folder:

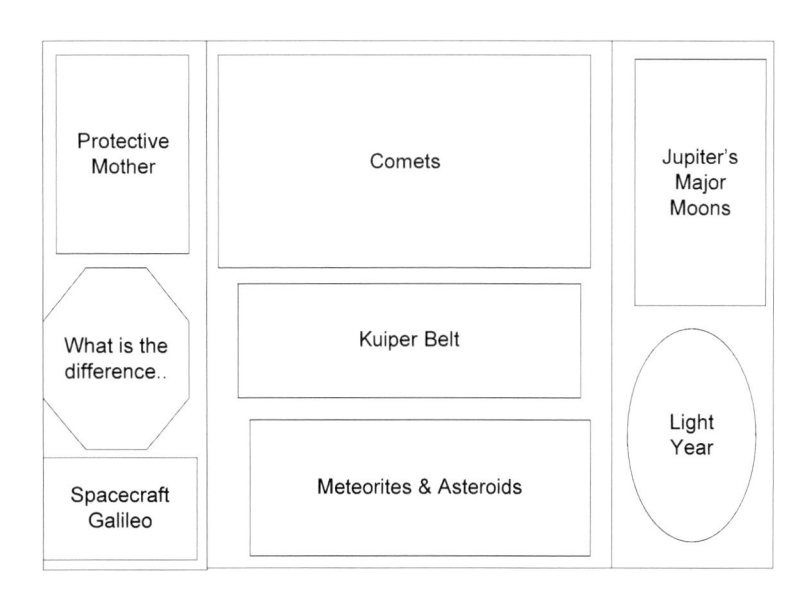

Protective Mother	Comets	Jupiter's Major Moons
What is the difference..	Kuiper Belt	Light Year
Spacecraft Galileo	Meteorites & Asteroids	

Inside of 3rd Folder:

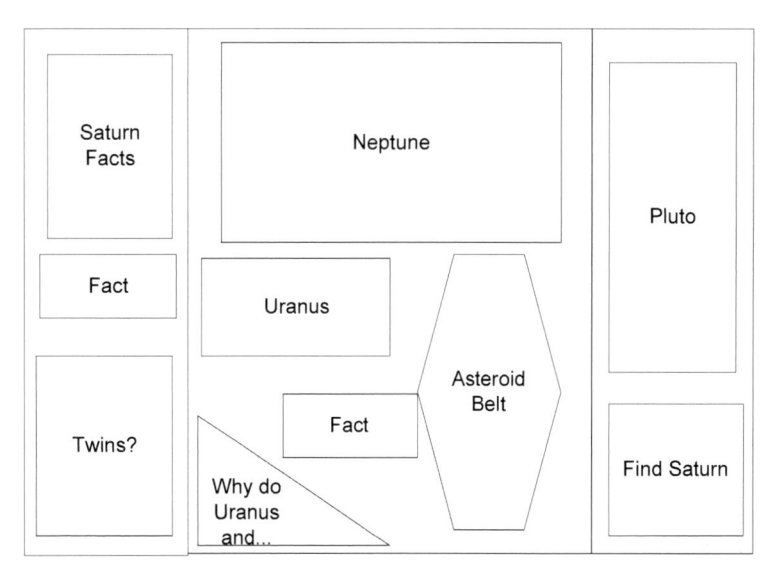

Saturn Facts	Neptune	Pluto
Fact	Uranus	
Twins?	Fact / Asteroid Belt	Find Saturn
	Why do Uranus and...	

Inside of 4th Folder:

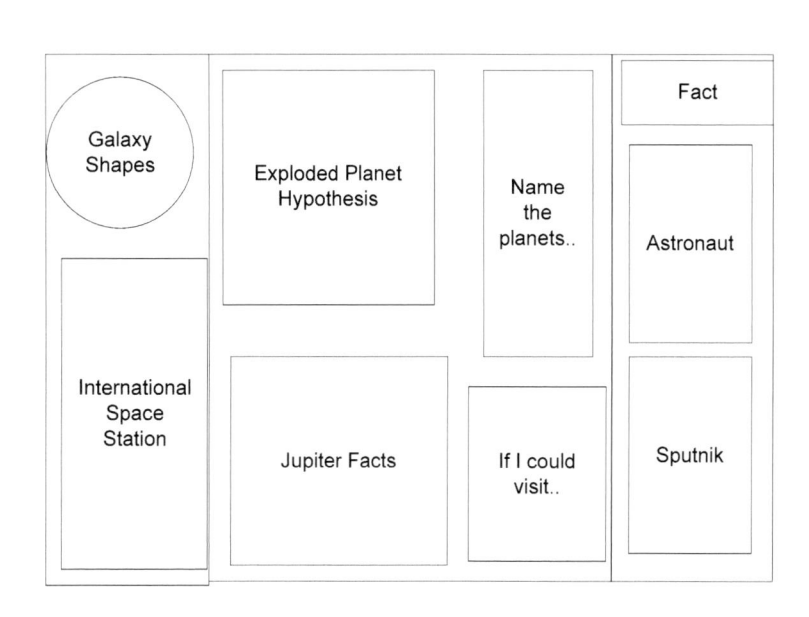

Galaxy Shapes	Exploded Planet Hypothesis	Name the planets..	Fact
			Astronaut
International Space Station	Jupiter Facts	If I could visit..	Sputnik

Below are pictures of a completed lapbook!!! This should help in figuring out how to assemble the booklets and then how to put it all together!

Completed
Lapbook

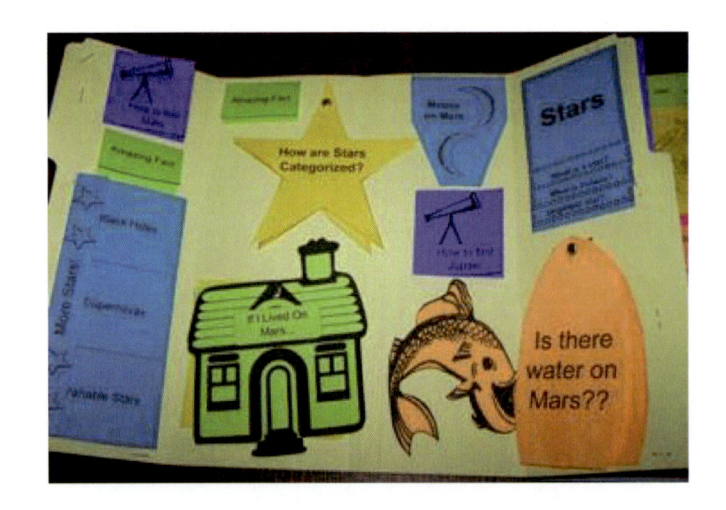

1ˢᵗ Folder

Continue ON…….. ➔

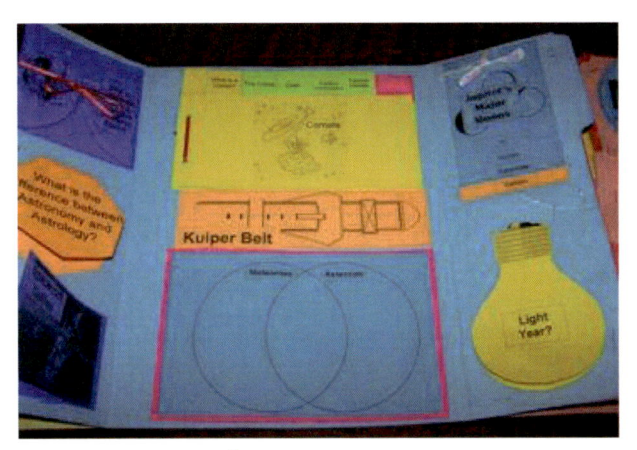

2nd Folder

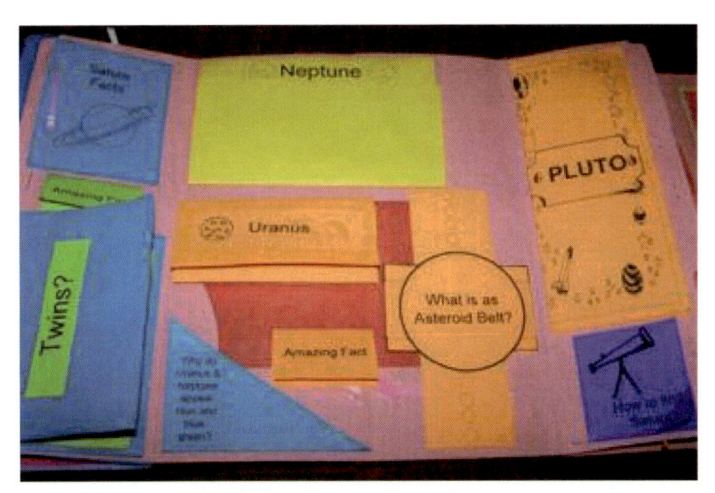

3rd Folder

4th Folder

Exploring Creation with Astronomy
Lessons 7 - 14
Lapbook Student Instruction Guide

Decorate the outside/ cover of your lapbook. The cover of your lapbook has purposely been left blank so that you may decorate it in any style you choose. You may draw, paint, glue pictures, etc. You may print coloring sheets and glue them to the front. The ideas for the front cover are endless. Choose a topic or art project from this unit and HAVE FUN making it your own personalized lapbook!

Lesson 7:
o "If I Lived On Mars" Booklet: Have you ever thought about going to another planet? Mars is a planet where some scientists would like to live. However, it would take a lot more work to move there and live there than you might think. In this booklet, write (or draw/paste pictures) about what you would have to take with you, how the land would look, where you would live, the atmosphere, how much you would weigh, time, days, seasons, and temperatures. You may also want to color the house on the front of the booklet. Templates on pages 21-24.
o Water On Mars Booklet: Do you think there is water on Mars? Color the fish, and answer his question. Template is on page 27.
o Moons On Mars Booklet: Did you know there are moons on Mars? Tell about them here. Template on page 25.
o How to Find Mars Booklet: Have you ever seen Mars? How do you find it? Write your answer here. Template on page 26.

Lesson 8:
o Comets Booklet: Have you ever seen a comet? Use this booklet to write about them. Discuss what a comet is, the coma, its orbit, creation confirmation, famous comets, and anything else interesting that you learned about comets in this lesson. Templates on pages 28-31.
o Meteorites & Asteroids Venn Diagram Booklet: Use this venn diagram to compare meteorites and asteroids. Template on page 35.
o Asteroid Belt: What is an asteroid belt? Can you wear it around your waist? I don't think so! Write about it here in this booklet. Template on page 32.
o Exploded Planet Hypothesis: Do you know what a hypothesis is? Did you know that some people have the hypothesis that there was once a planet between Jupiter and Mars? Explain this theory here. You may also want to color the front of the booklet. Templates on pages 33-34.

Lesson 9:
o Protective Mother Booklet: Do you know any "protective mothers"? Did you know that the Earth has one too? Explain that here. Template on page 39.
o Jupiter Facts Booklet: Jupiter is an amazing planet. There are so many interesting facts. Open up each of the booklets inside this pocket, and tell about Jupiter's red spot, little sun, rings, rotation & revolution, and why it is called a gas giant. Templates on pages 42-47.
o Jupiter's Major Moons Booklet: Did you know that Jupiter has more than 60 moons?? Write what you have learned about Jupiter's 4 major moons here. Templates on pages 37-38.
o How to Find Jupiter Booklet: Did you know that you can see Jupiter if you know where and when to look? Explain how to find Jupiter here. Template on page 40.
o Spacecraft Galileo Booklet: Have you heard of spacecrafts with no astronauts on them? Tell about this one here. Template on page 41.

Continue ON........ ⟶

Lesson 10:
o   Saturn & Jupiter "Twins?" Booklet: Do you know anyone who has a twin? Saturn and Jupiter are a lot like twins. Use this venn diagram to explain the similarities and differences between these planets. Template on page 50.
o   Saturn Facts Booklet: Saturn has a day of the week named after it...do you know what it is? Inside this booklet, explain what you know about Saturn's rotation, moons, ring system, and the Cassini Mission. Templates on pages 48-49.
o   How to Find Saturn Booklet: Saturn is not hard to find in the night sky. You can even see its rings! Explain how to find it here. Template is on page 47.

Lesson 11:
o   "Why do Uranus & Neptune appear blue and blue-green" Booklet: These 2 planets have something unique about their atmosphere that makes them look this color. Do you know what it is? Template on page 55.
o   Uranus Facts Booklet: How much have you learned about this planet? See if you remember this information, and record your answers. Template on page 51.
o   Neptune Facts Booklet: Neptune is a beautiful blue planet. It has some interesting characteristics. Inside these booklets, tell about them. Templates on page 52-54.

Lesson 12:
o   Kuiper Belt Booklet: This is another belt to study! It is similar, yet different than the other belt that you studied, the asteroid belt. Tell about it here. Template on page 56.
o   Pluto Facts Booklet: Pluto is very small, and maybe not even a planet at all. After you study about it, see what you think….is it a planet? Templates on pages 57-58.

Lesson 13:
o   Stars Booklet: Stars produce a lot of energy...energy that makes them so bright that we can see them. Have you ever stood outside at night and looked up at them? What is the brightest one? See if you can answer these questions. Template on page 59.
o   More Stars Booklet: This booklet is about special types of stars….see if you remember what they are! Template on page 60.
o   How are Stars Categorized Booklet: Scientists can tell a lot about a star by certain characteristics. Do you know what they are? Templates on pages 63-64.
o   Galaxy Shapes Booklet: Stars are found in groups called galaxies. These galaxies come in 4 different shapes. Do you know the name of our galaxy? Tell about these galaxies here. Template on page 61.
o   "What is the difference…" Booklet: You have been learning about "astronomy", but have you ever heard of "astrology"? These are 2 very different terms. Explain the difference here. Template on page 62.
o   Light Year Booklet: Do you know what a light year is? Does it have anything to do with a light bulb? Explain the definition here. Template on page 65.

Lesson 14:
o   Sputnik & Space Race Booklet: Have you ever heard of "Sputnik"? It sounds like a funny word...doesn't it? What is it? And what is the "space race"? Write your answers here. Template on page 66.
o   International Space Station Booklet: Why was this the "best space station ever built"? Tell what you know about it here...and color its picture on the front. Templates on pages 67-68.
o   Becoming a NASA Astronaut: Have you ever dreamed of becoming an astronaut? How would you go about it? Template on page 69.
o   "If I could Visit…" Booklet: Would you like to visit another planet? If you could visit any planet, which one would you want to visit and why? Explain here. Template on page 70.
o   Name the Planets Booklet: Now that you have studied all of the planets, can you name them in order? Try to do that here. Template on page 71.

Exploring Creation With Astronomy
Lessons 7- 14
Lapbook Assembly Guide

Inside of 1st Folder:

1. Saturn Facts Booklet: Cut out each page along outer black lines. Stack them so that the title is on the top. Punch 2 holes in the far left side of the stack. Secure by tying a ribbon through the holes.
2. Amazing Fact Booklet (There are 2 of these in this folder): Cut out along the outer black lines, and fold in the center, so that the title is on the front.
3. Twins Booklet: Cut out this venn diagram booklet along the outer black line edges. Glue to a slightly larger piece of paper of a different color, creating a small border around the edges. Cut out the title/text box also. Fold the booklet in half, vertically. Glue the text box to one side. Secure only one side of the booklet to the folder. See picture below:

Inside of booklet: Outside of booklet:

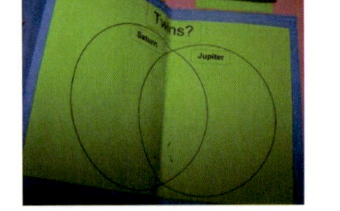

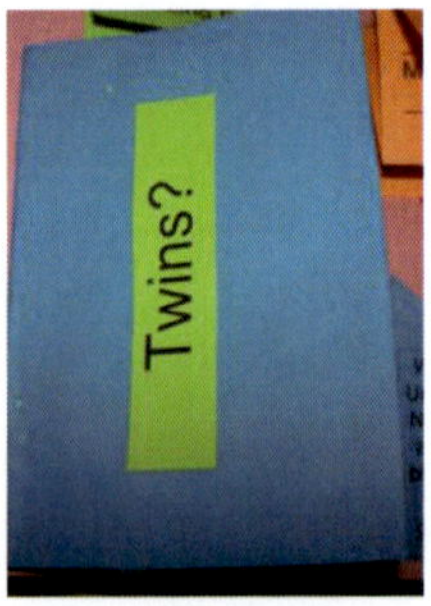

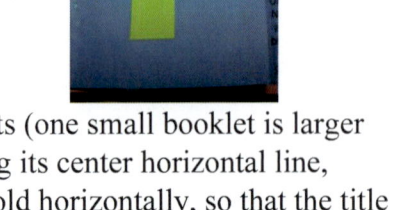

4. Neptune Booklet: Cut out the background page and the 5 small booklets (one small booklet is larger than the others) along the outer black line edges. Fold each small booklet along its center horizontal line, keeping the title on the front. Glue the booklets to the background page, and fold horizontally, so that the title on the background page is still visible. See picture:

Inside of booklet: Outside of booklet:

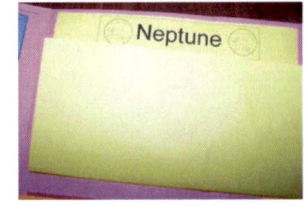

5. Uranus Booklet: Cut out along the outer black line edges of the booklet and the title box. Fold accordion-style, so that the words are all on the inside, and then glue the title box to the top. See picture:

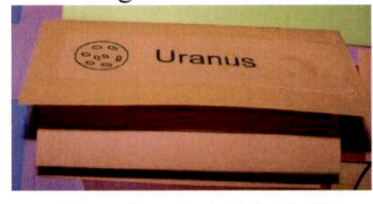

6. "Why do Uranus & Neptune…" Booklet: Cut out along the outer black line edges, and then fold along the center line so that the title is on the front.

Continue ON…….. ➡

7. Asteroid Belt Booklet: Cut out along the outer black line edges of the booklet. Then, fold along the center vertical line, so that the title is on the front.

8. Pluto Booklet: This booklet is designed to be printed out double-sided. If your printer will not do this, just print out 2 pages, and glue them back to back. Then, tri-fold the booklet so that there is a blank space on the back, and the title is on the front.

9. Find Saturn Booklet: Cut out along the outer black line edges of the booklet. Fold along the center vertical line so that the title is on the front.

Inside of 2nd t Folder:

1. Find Mars Booklet: Cut out along the outer black line edges of the booklet. Fold along the center vertical line so that the title is on the front.

2. Amazing Fact Booklets (2): Cut out these booklets along the outer black line edges. Fold along the center horizontal line so that the titles are on the front.

3. More Stars Booklet: Cut out along the outer black line edges of the booklet. Fold along the vertical line closest to the center, so that the words are on the front. Now, fold along the other vertical line (with the title on it) and it will fold over the edges of the words. Now, cut along the horizontal lines between the words. The title portion of the booklet will keep the larger vocabulary folds closed. See picture:

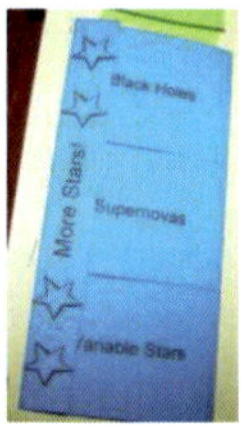

4. Moons on Mars Booklet: Cut out along the black line edges of the booklet. Fold along the middle line so that the title is on the front.

5. Find Jupiter Booklet: Cut out along the outer black line edges of the booklet, and then fold along the center line so that the title is on the front.

6. How are Stars Categorized Booklet: Cut out each of the stars. Stack them so that the title is on the top. Punch one hole through the top point section, and secure with a metal brad fastener.

7. If I Lived On Mars Booklet: Cut out each page along the black line edges. Stack the pages together, and keep the title/graphic page on top. Punch a hole through the center of the top of the house, and secure with a metal brad fastener. See picture:

8. Water on Mars Booklet: Cut out along the outer black line edges of both pieces of the booklet. Place the oval shape with the title on it on top of the other one. Punch a hole through the top of the 2 pages, and secure with a metal brad fastener. This booklet is glued into the folder along the fold, where the fish is on one side of the fold and the booklet is on the other side of the fold. See picture:

9. Stars Booklet: Cut out each page along the outer black line edges. Stack them so that the title is on top, and each page gets progressively longer toward the back page. Secure at the top with 2 staples. See picture:

Inside of 3rd Folder:

1. Protective Mother Booklet: Cut out along the outer black line edges. Fold along the 2 vertical lines, so that the words are on the outside, and the rounded edges almost touch in the center. Punch a hole through each of the two circles where the edges almost touch. Secure with a ribbon or yarn, tied loosely in a bow. See picture:

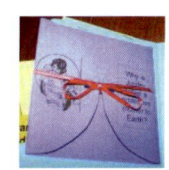

Continue ON……..

2. Difference between...Booklet: Cut out along the outer black line edges. Fold along the center line so that the title is on the front.

3. Spacecraft Galileo Booklet: Cut out along the outer black line edges. Fold along the vertical line, so that the title is on the front. This booklet folds from the opposite direction that you would expect.

4. Comets Booklet: Cut out each page of the booklet. Note that each page has a longer tab at the top. Stack the pages together, so that the tabs get longer from front to back. Punch 2 holes on the far left side of the booklet. Secure with a ribbon or yarn. See picture:

5. Kuiper Belt Booklet: Cut out along the outer black line edges. Fold along the center line so that the title is on the front.

6. Meteorites & Asteroids Booklet: Cut out the booklet along the outer black line edges. Glue onto a piece of paper of a different color, creating a small border around the edges.

7. Jupiter's Major Moons Booklet: This booklet is cut out and assembled exactly like the "Stars" booklet in the 2nd folder. The only difference is that this one is secured at the top by punching 2 holes and placing a ribbon through them.

8. Light Year Booklet: Cut out along the outer black line edges of the booklet. Fold along the center line so that the title is on the front.

Inside of 4th Folder:

1. Galaxy Shapes Booklet: Cut out along the outer black line edges of both circles. Cut out the wedge on the title page. Place the title page on top of the other one, and punch a hole through the center of the circles, and secure loosely with a metal brad fastener (loosely so that the top circle will turn). See picture:

2. International Space Station Booklet: Cut out each of the pages of the booklet, and stack on top of each other, with the title/graphic page on top. Punch 2 holes along the left side, and secure with a ribbon or yarn.

3. Exploded Planet Hypothesis Booklet: Cut out the ringed planet, and cut it in half along the dotted line. Cut out the booklet along the outer black line edges also. Fold the booklet along its 2 dotted vertical lines, so that the edges almost meet in the front (circle on the front). Glue one side of the planet to each where the edges almost meet. See picture:

4. Jupiter Facts Booklet: There are 6 pieces to this booklet. They are labeled "Jupiter Facts," "Little Sun," "Rotation & Revolution," "Great Red Spot," "Gas Giant," and "Jupiter's Rings." Cut all of them out along the outer black line edges. On the one that says "Jupiter Facts," fold along the vertical line, so that the title is on the front. Glue around the bottom and right edges, creating a pocket. All of the other booklets that go inside the pocket are folded along their center lines, so that the titles are on the front. Now place all of the 5 booklets into the pocket. See picture:

Continue ON........

5. Name the Planets Booklet: Cut out along the outer black line edges of the booklet. Glue to a slightly larger piece of paper of a different color, so that you create a small border around the edges.
6. If I could visit...Booklet: Cut out along the outer black line edges of the booklet. Glue to a slightly larger piece of paper of a different color, so that you create a small border around the edges.
7. Amazing Fact Booklet: Cut out along the outer black line edges of the booklet, and fold along the center line so that the title is on the front.
8. Becoming a NASA Astronaut Booklet: Cut out along the edges of both spaceships. Place the one with the title on top, and punch a hole through the bottom center. Secure with a metal brad fastener.
9. Sputnik & Space Race Booklet: Cut out along the outer black line edges. Accordion-fold along the vertical lines, so that the title is on the front.

Exploring Creation With Astronomy
Lessons 7 – 14 Lapbook
Teacher's Guide

Here, you'll find information to supplement your study. Jeannie Fulbright's book is so wonderfully filled with knowledge and wisdom. All of the information needed to complete all of the booklets can be found on the pages of her book. Below, I will tell you which pages hold specific "answers." Also, you'll find many other sites listed, where you may want to go for extra information, coloring pages, games, crafts, and ideas to extend your study.

I have been questioned as to <u>why I merely give you the page numbers for the answers instead of the answers themselves</u>. If I were to give you ONLY the answers, then there would be no need for you to have Jeannie's awesome book...right? Also, this will require the parent to actually read the book as well, which was Jeannie's intent from the beginning. So, I hope that you understand my decision to not "just give the answers." It really is a calculated plan on mine and Jeannie's part.

Lesson 7:
- "If I Lived On Mars" Booklet: Answers found throughout Lesson 7
- Water On Mars Booklet: Answers found on pages 84-85
- How to Find Mars Booklet: Answer found on page 85

Additional Resources for Lesson 7:
* Lots of fun & games about Mars: http://mars.jpl.nasa.gov/funzone_flash.html
* Great resources for information about Mars: http://www.kidscosmos.org/kid-stuff/mars-facts.html

Lesson 8:
- Comets Booklet: Answers found on pages 90-94
- Meteorites & Asteroids Venn Diagram Booklet: Answers found on pages 95-96
- Asteroid Belt: Answer found on page 96-99
- Exploded Planet Hypothesis: Answers found on pages 97-99

Additional Resources for Lesson 8:
* Website with comet information: http://library.thinkquest.org/3645/comets.html
* Asteroids information: http://www.worldalmanacforkids.com/WAKI-ViewArticle.aspx?pin=wwwwak-182&article_id=535&chapter_id=12&chapter_title=Science&article_title=Asteroid
* Asteroid Belt information: http://starchild.gsfc.nasa.gov/docs/StarChild/solar_system_level1/asteroids.html

Lesson 9:
- Protective Mother Booklet: Answer found on page 104
- Jupiter Facts Booklet: Answers found on pages 104-106
- Jupiter's Major Moons Booklet: Answers found on pages 106-108
- How to Find Jupiter Booklet: Answer found on page 110
- Spacecraft Galileo Booklet: Answer found on pages 108-109

Additional Resources for Lesson 9:
* Website for "Galileo's Probe Into Jupiter" - http://quest.arc.nasa.gov/galileo/edbrief/galileotoc.html
* Lots of information about Jupiter: http://www.happynews.com/living/space/jupiter-information-kids.htm
* Great site about Jupiter: http://www.kidsastronomy.com/jupiter.htm

Lesson 10:
- Saturn & Jupiter "Twins"? Booklet: Answers found on pages 114-115
- Saturn Facts Booklet: Answers found on pages 115-117
- How to Find Saturn Booklet: Answer found on page 118

Continue ON…….. ⟶

Additional Resources for Lesson 10:
* Saturn Information: http://www.kidsastronomy.com/saturn.htm
* EXCELLENT site with lots of information and activities about Saturn: http://saturn.jpl.nasa.gov/kids/index.cfm

Lesson 11:
- "Why do Uranus & Neptune appear blue and blue-green" Booklet: Answer found on page 122
- Uranus Facts Booklet: Answers found on pages 122-124
- Neptune Facts Booklet: Answers found on pages 125-127

Additional Resources for Lesson 11:
* Uranus information from an astronomer: http://coolcosmos.ipac.caltech.edu//cosmic_kids/AskKids/uranus.shtml
* Fun Neptune information: http://www.lunaroutpost.com/library/astronomy_kids/kid_neptune.htm
* More fun Neptune information: http://kids.nineplanets.org/neptune.htm

Lesson 12:
- Kuiper Belt Booklet: Answer found on page 132
- Pluto Facts Booklet: Answers found on pages 133-138

Additional Resources for Lesson 12:
* Kuiper and Asteroid Belts information: http://www.kidsastronomy.com/academy/lesson110_assignment1_6.htm
* More great information: http://www.sciencenewsforkids.org/articles/20040407/Feature1.asp
* Pluto facts: http://www.worldalmanacforkids.com/explore/space/pluto.html
* Is Pluto a planet: http://news.nationalgeographic.com/news/2006/08/060824-pluto-planet.html

Lesson 13:
- Stars Booklet: Answers found on pages 142-144
- More Stars Booklet: Answers found on pages 144-146
- How are Stars Categorized Booklet: Answers found on pages 147-148
- Galaxy Shapes Booklet: Answers found on pages 150-151
- "What is the difference…" Booklet: Answer found on page 154

Additional Resources for Lesson 13:
* All about stars: http://www.kidsastronomy.com/stars.htm
* Fun and games about stars: http://starchild.gsfc.nasa.gov/docs/StarChild/StarChild.html
* Galaxy craft: http://spaceplace.nasa.gov/en/kids/galex/art.shtml

Lesson 14:
- Sputnik & Space Race Booklet: Answers found on pages 161-163
- International Space Station Booklet: Answers found on pages 163-169
- Becoming a NASA Astronaut: Answers found on pages 168-169

Additional Resources for Lesson 14:
* Sputnik information and video: http://news.bbc.co.uk/onthisday/hi/dates/stories/october/4/newsid_2685000/2685115.stm
* How to become an astronaut: http://people.howstuffworks.com/question534.htm

Continue ON……..

Additional Resources for ALL LESSONS:
* Lots of great information on all of the planets:
  http://www.kidsastronomy.com/
* More great information, sponsored by NASA:
  http://image.gsfc.nasa.gov/poetry/ask/askmag.html
* Lots of great information and games/activities for kids:
  http://starchild.gsfc.nasa.gov/docs/StarChild/StarChild.html
* Really fun games from NASA:
  http://www.nasa.gov/audience/forkids/kidsclub/flash/index.html
* Yet another wonderful site: http://www.esa.int/esaKIDSen/index.html

If I Lived On Mars....

Lesson 7

What I would have to take...

Continue ON........ ➡

Where I would live

How the land would look...

Lesson 7

Continue ON…….. ➡

How much I would weigh

Atmosphere

Lesson 7

Continue ON…….. ➡

Time & Days...

Seasons & Temperatures

Lesson 7

Continue ON........ ➤

Moons on Mars

How to find Mars

Lesson 7

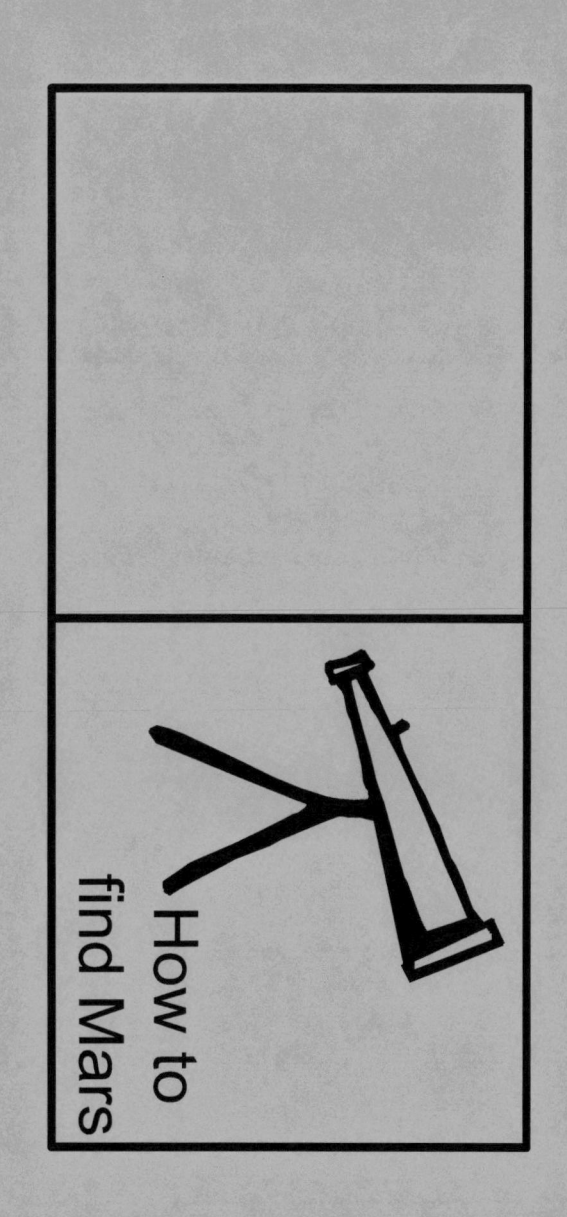

Is there water on Mars??

Lesson 8

Suggested Paper Color: Blue

Comets

What is a Comet?

Lesson 8   **Continue ON……..**

The Coma

Orbit

Lesson 8   Continue ON……..

Creation Confirmation

Famous Comets

Lesson 8   **Continue ON……..** ➡

Lesson 8

What is an Asteroid Belt?

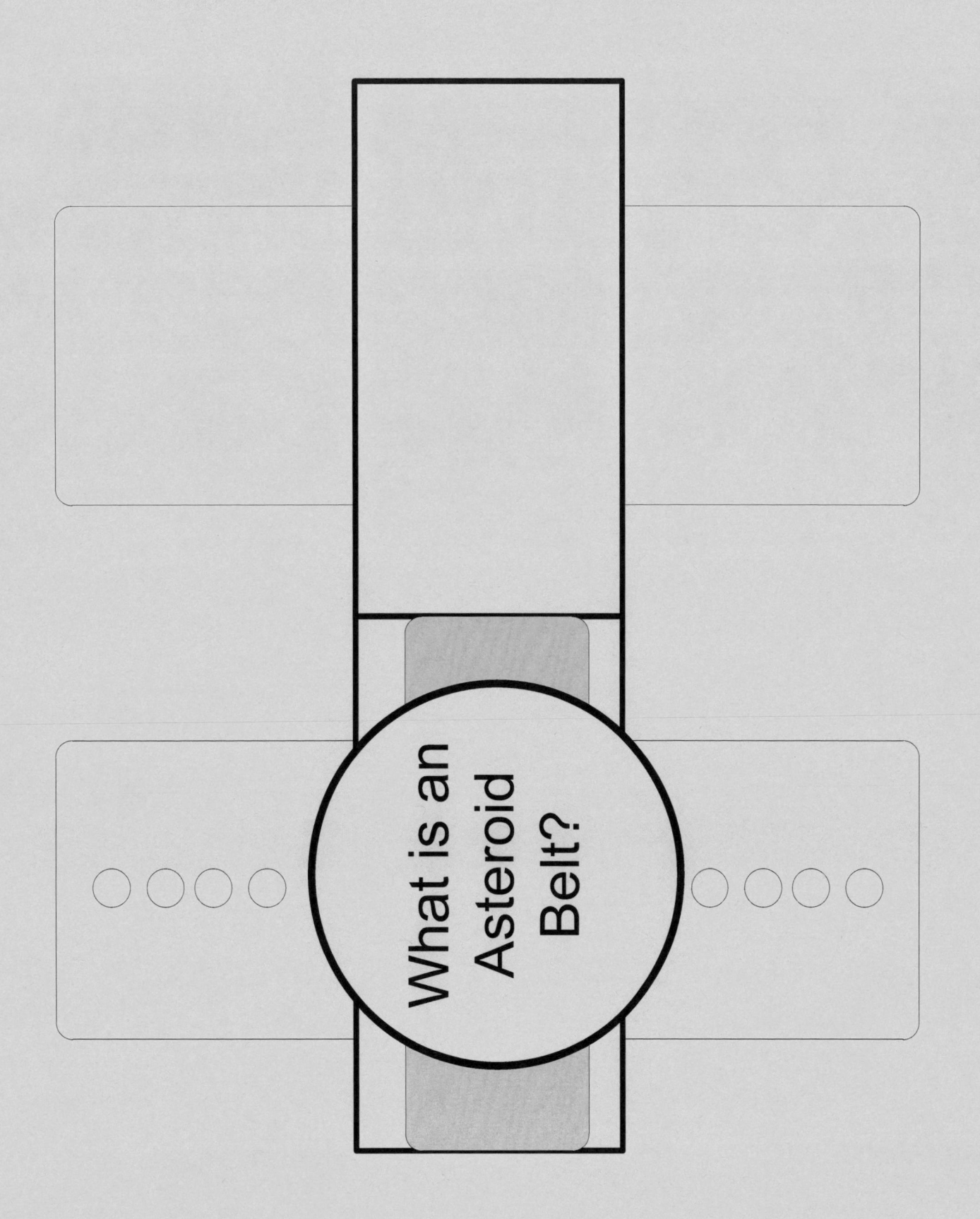

Lesson 8

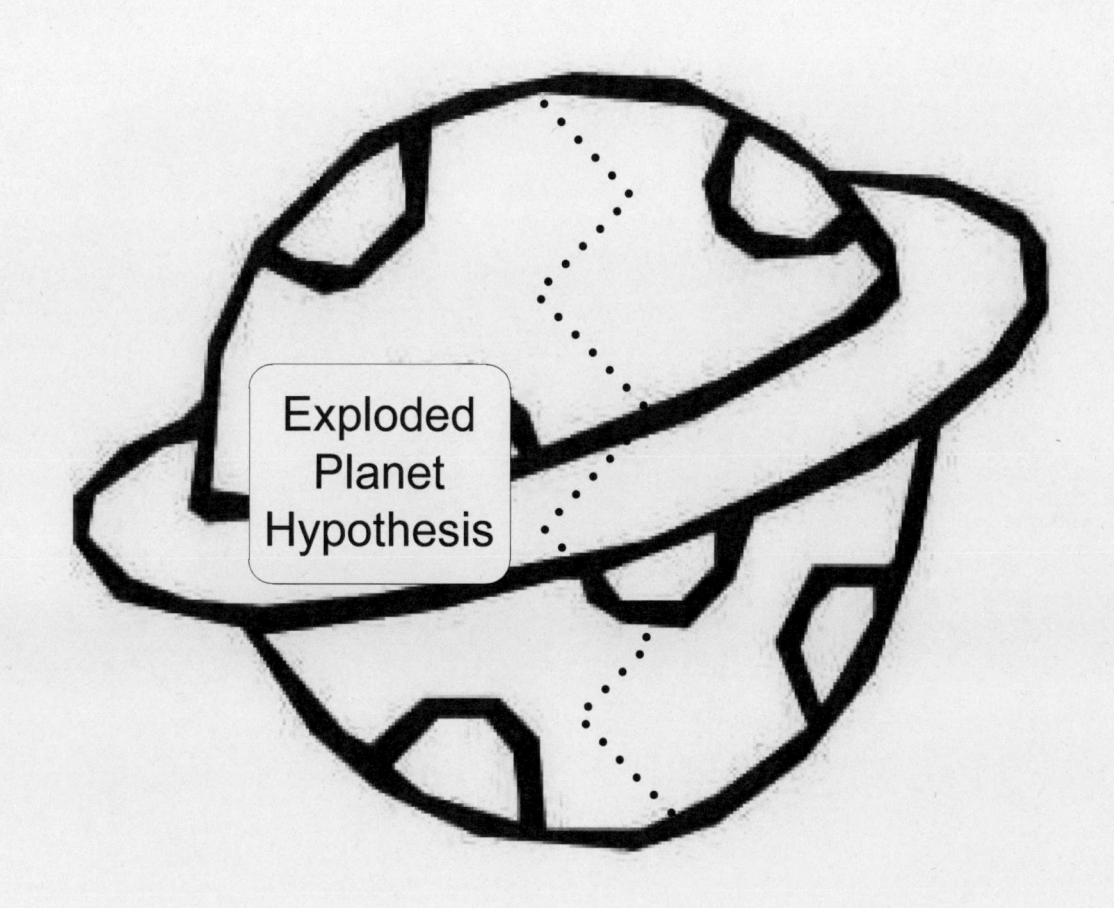

Exploded
Planet
Hypothesis

Lesson 8   Continue ON……..

Exploded Planet Hypothesis – Follow instructions to glue the earth on the previous page onto this booklet.

Lesson 8

Asteroids

Meteorites

Lesson 8

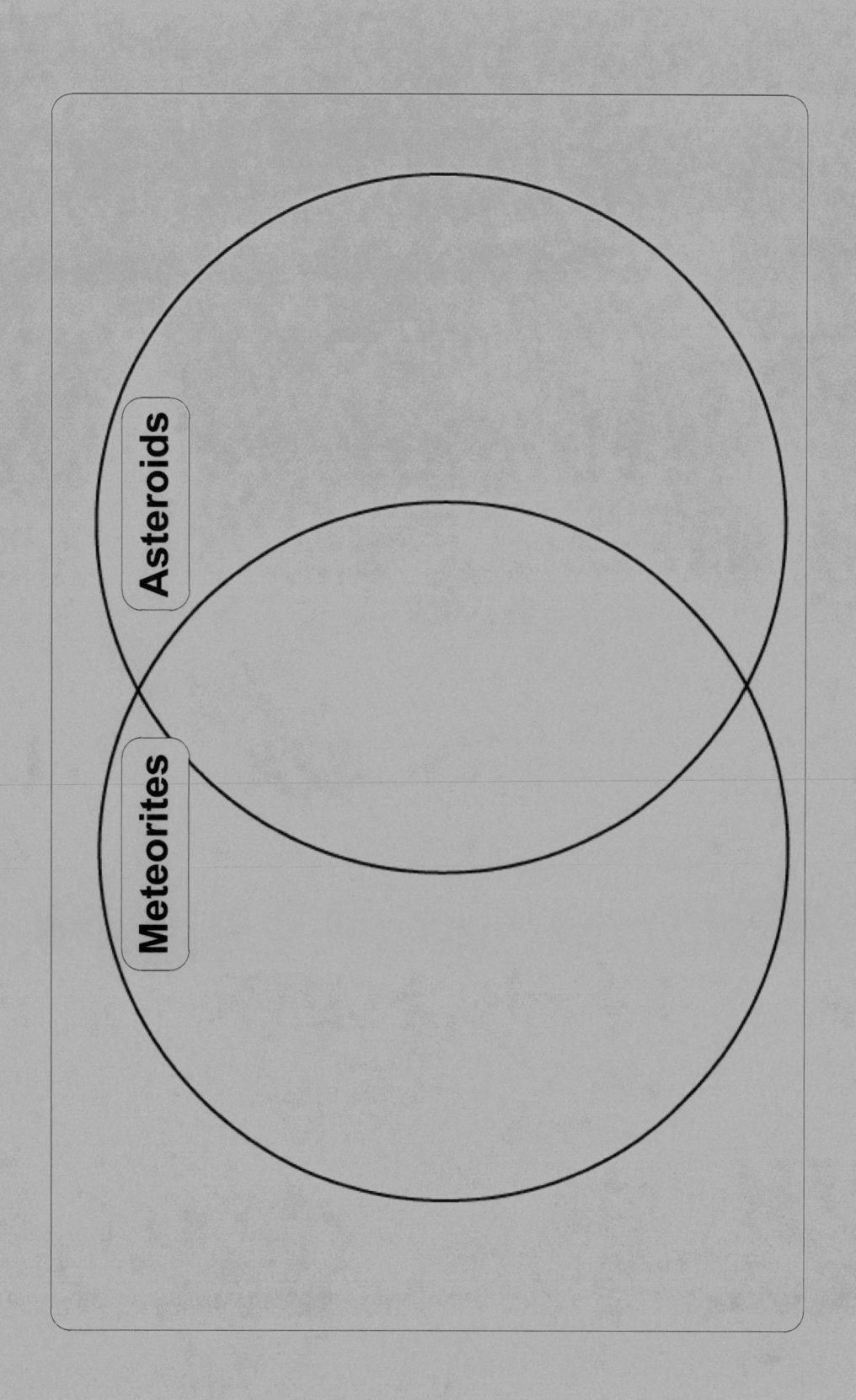

Jupiter's Major Moons

Io

Europa

Ganymede

Lesson 9   Continue ON……..

Callisto

Lesson 9

Why is Jupiter called a protective mother to Earth?

Lesson 9   **Suggested Paper Color: Green**

How to
find
Jupiter

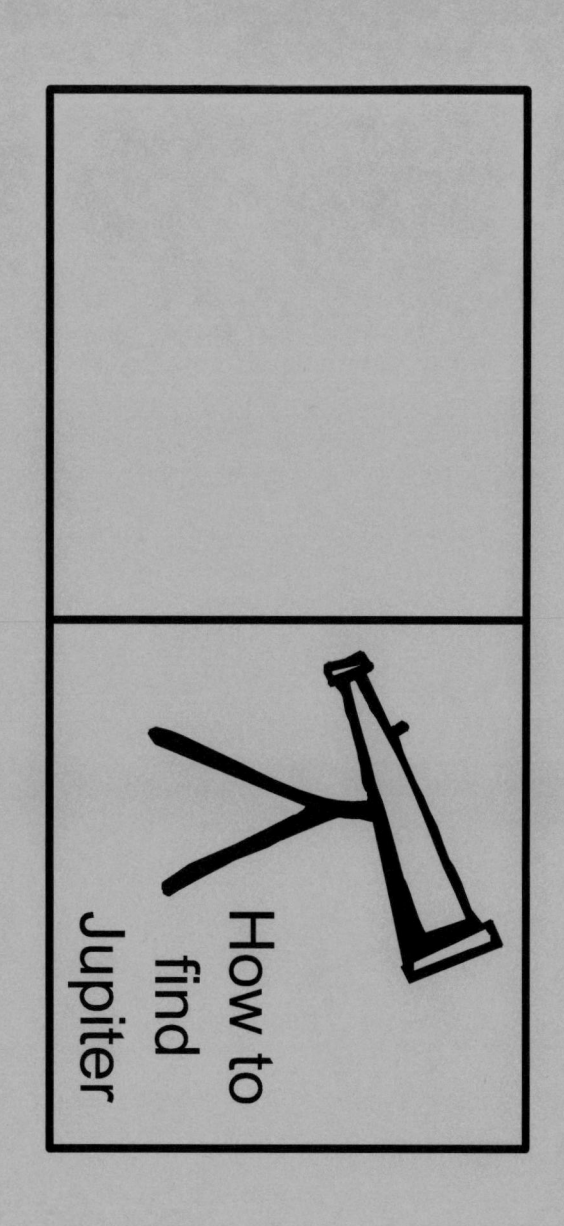

Lesson 9

Spacecraft Galileo

Lesson 9

Jupiter Facts

Lesson 9

Continue ON……..

Rotation
&
Revolution

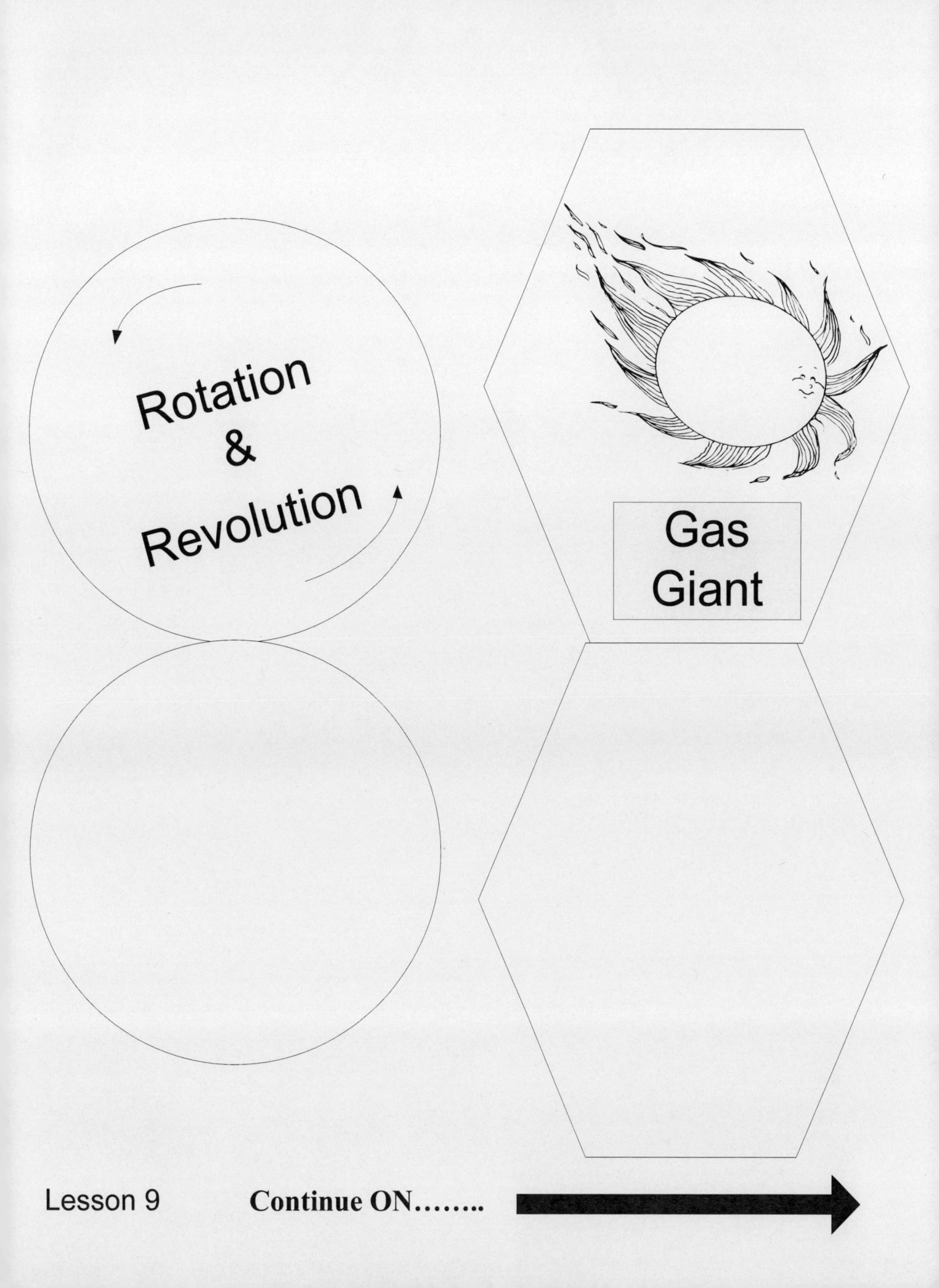

Gas
Giant

Lesson 9   Continue ON……..

Jupiter's Rings

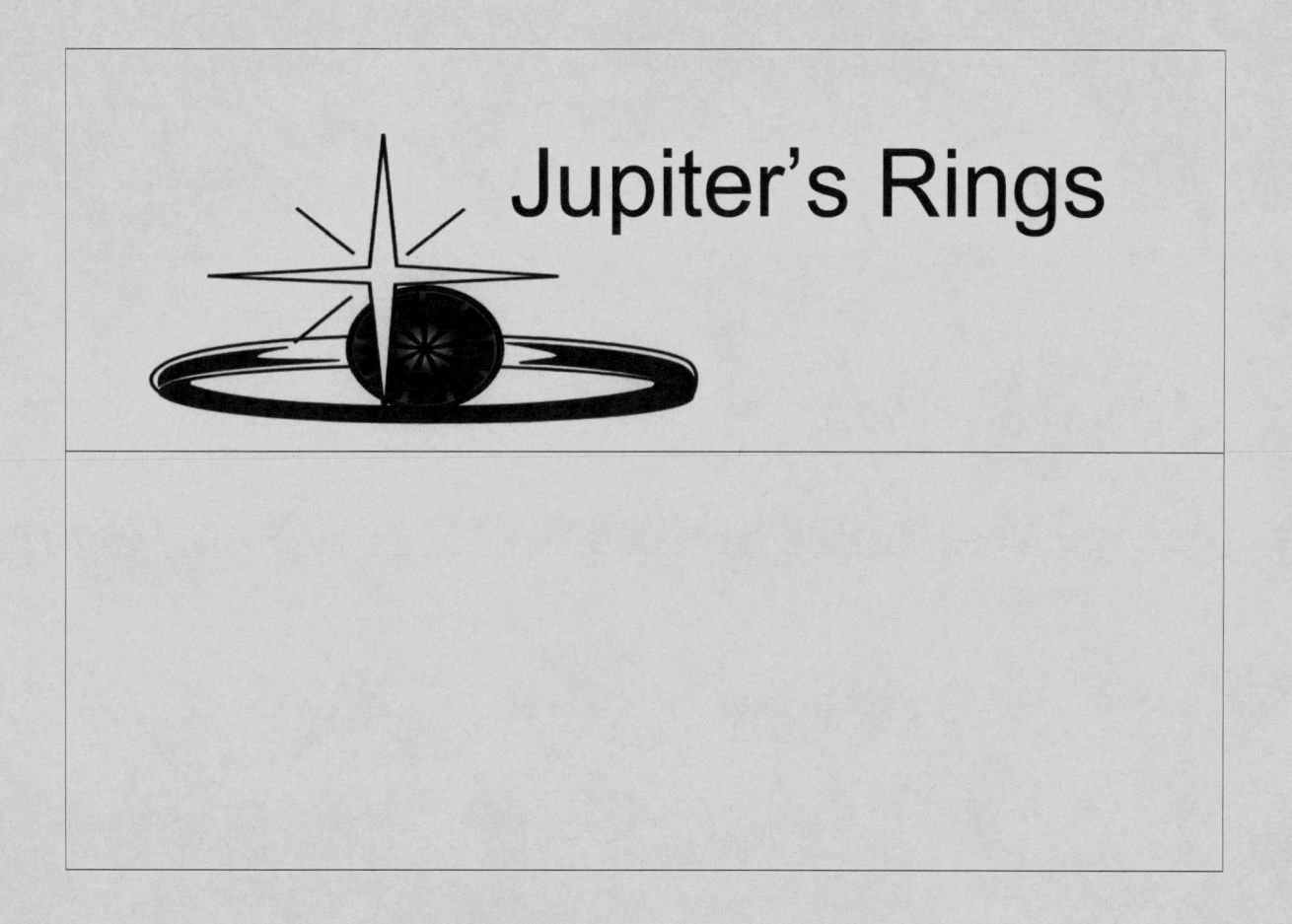

Lesson 9   **Continue ON……..**

Little Sun

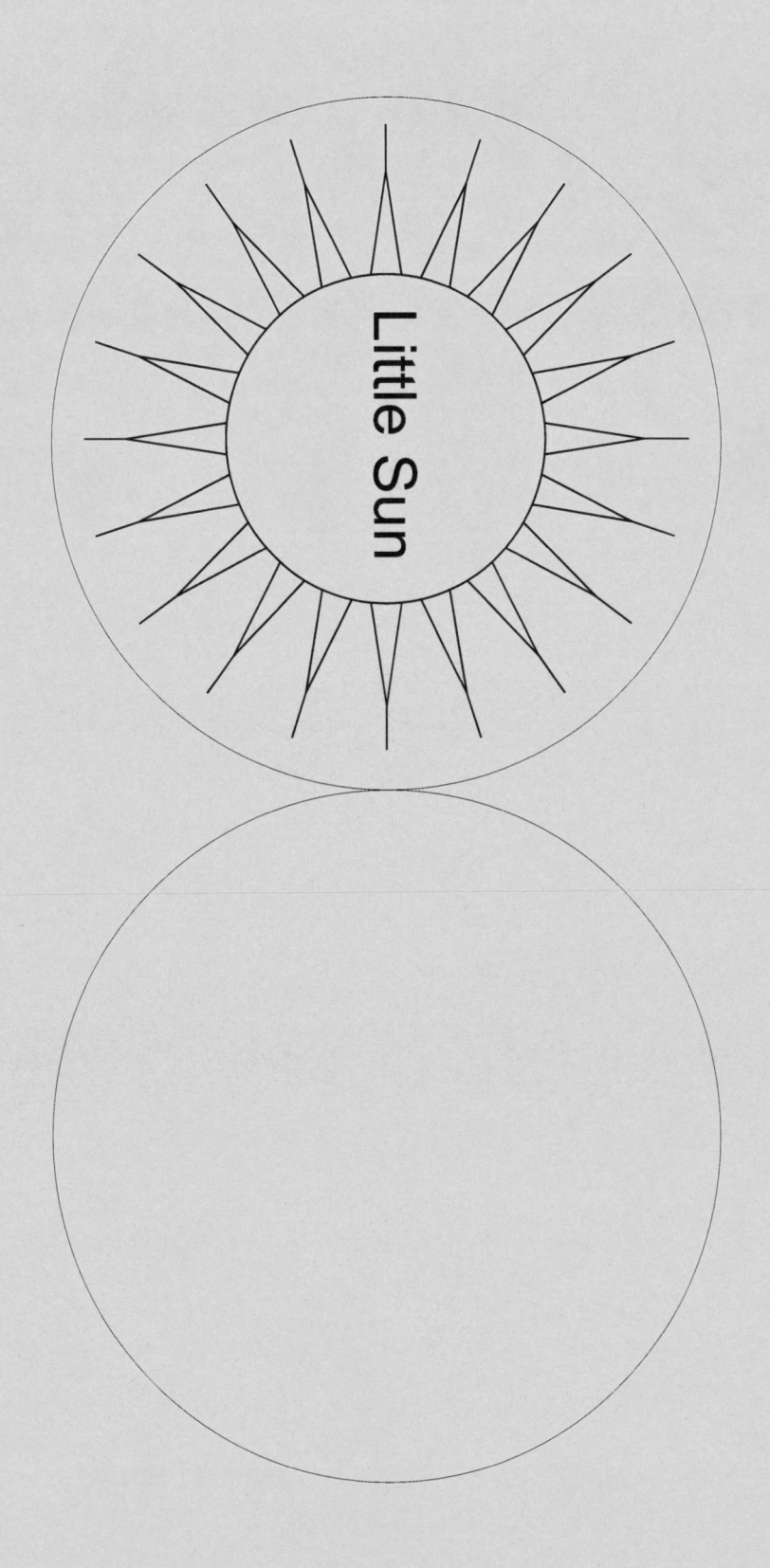

Suggested Paper Color: Yellow

Lesson 9   **Continue ON........** ➡

Great Red Spot

Suggested Paper Color: Red

Lesson 9

How to find Saturn

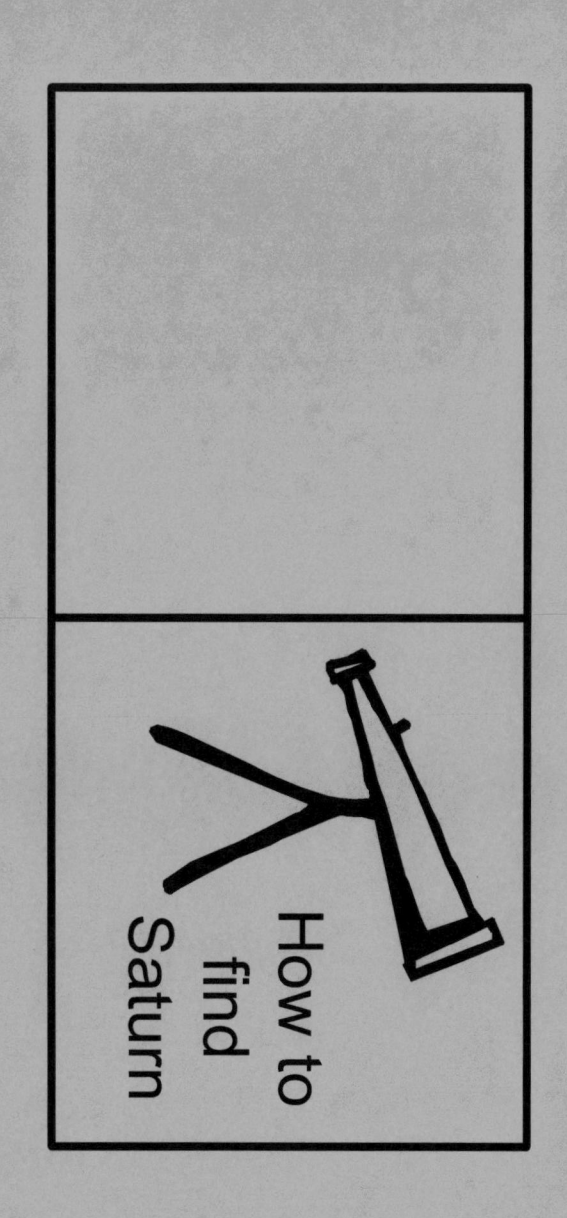

Lesson 10

Saturn Facts

Ring System

Rotation

Cassini Mission

Lesson 10   **Continue ON……..** ➡

Moons

Lesson 10

Twins?

Twins?

- Jupiter
- Saturn

Lesson 10

Position: _____
Temperature: _____
Rings: _____
_____

Rotation: _____
_____
Moons: _____
_____

When was Uranus discovered, and by whom? _____
_____
_____
_____

Orbit & Rotation: _____
_____
_____
_____

Uranus

Lesson 11

Neptune

Lesson 11   **Continue ON……..**

| Number ? | Rotation & Revolution | Moons |

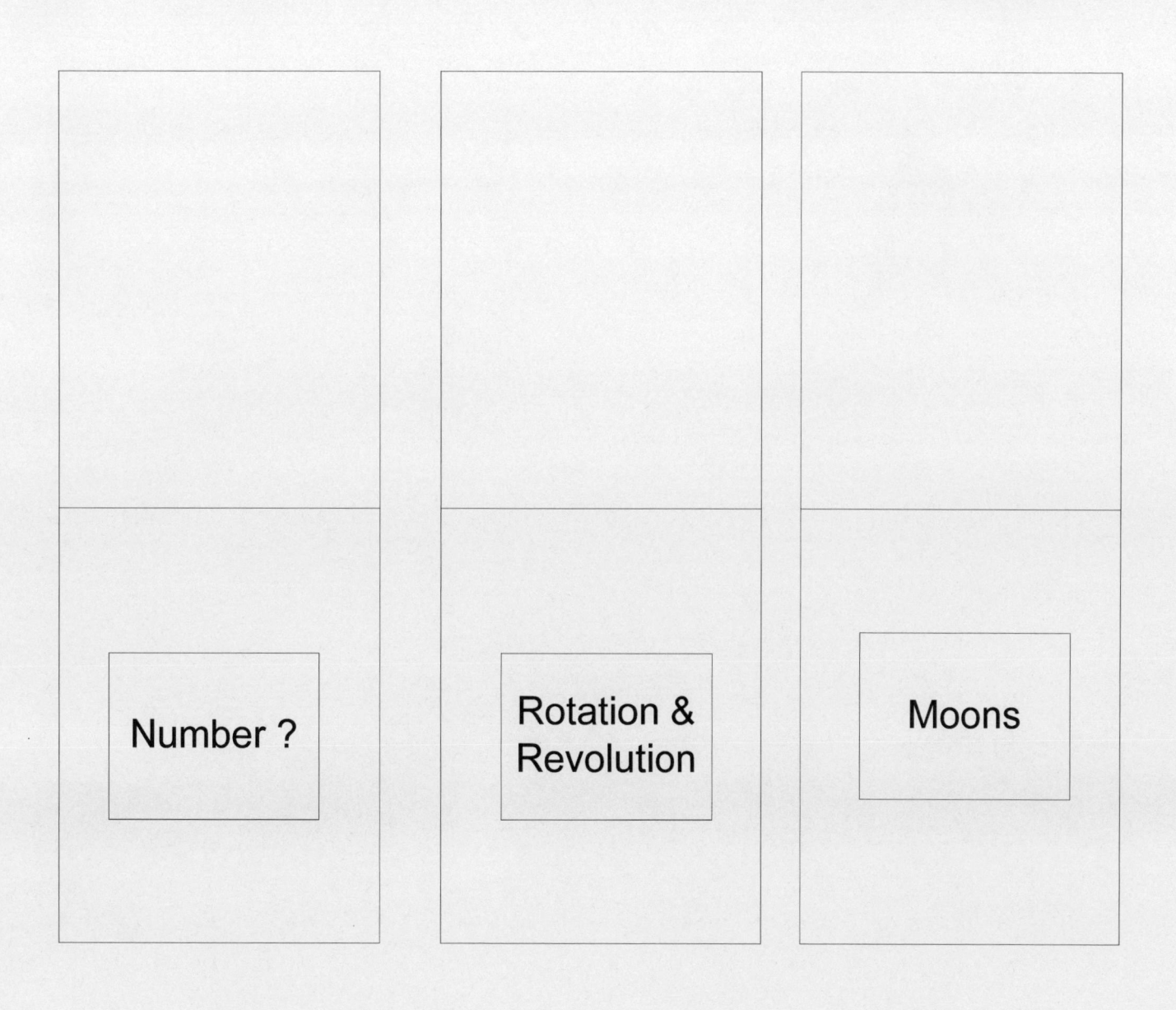

Lesson 11   **Continue ON……..**   ➡

Atmosphere

Discovery

Lesson 11

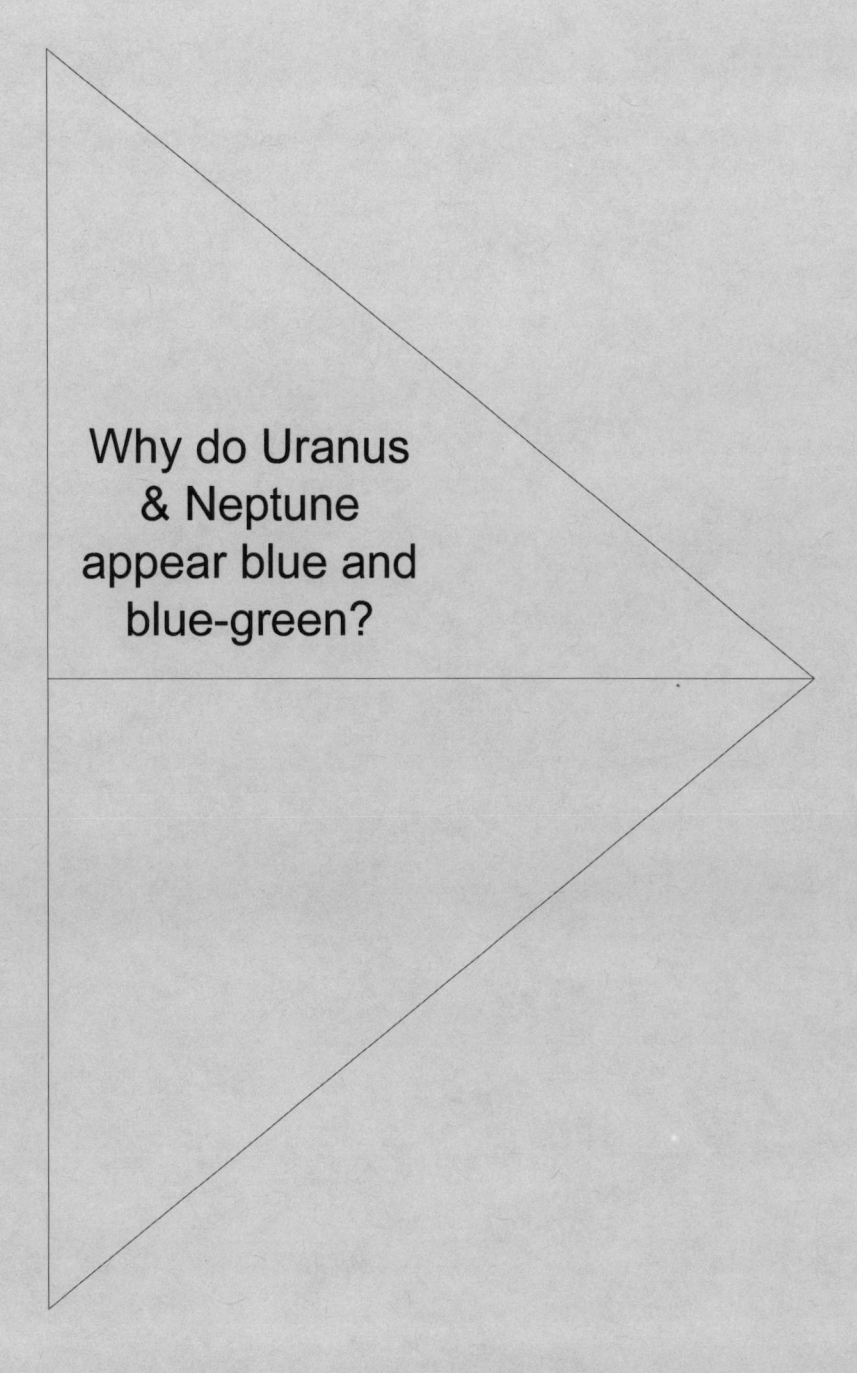

Why do Uranus
& Neptune
appear blue and
blue-green?

Suggested Paper Color: Blue

Lesson 11

Kuiper Belt

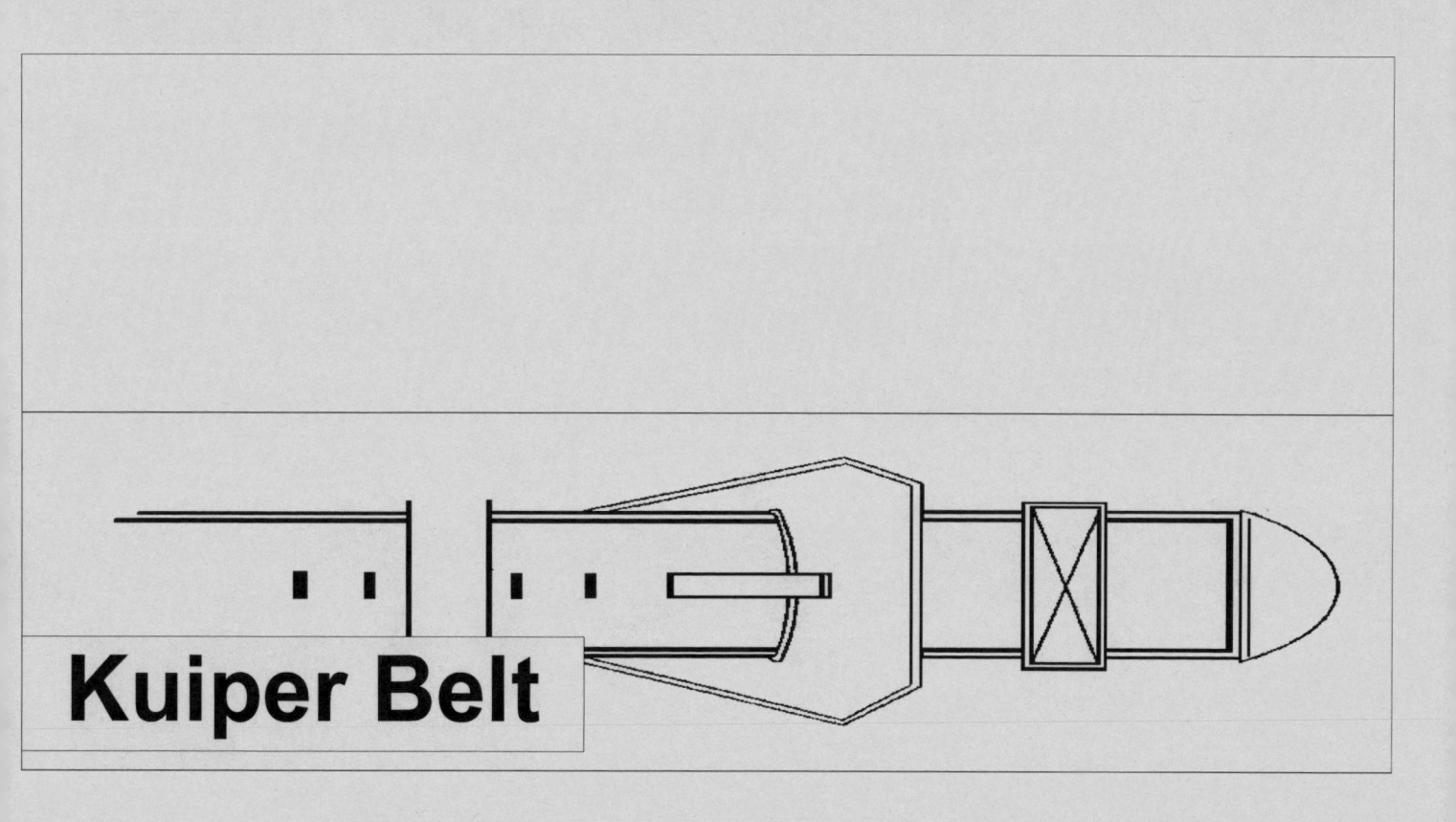

Lesson 12

PLUTO

Arguments "AGAINST"
Pluto being a planet:

What do YOU think?
Planet or Comet?

Planet Pluto or Comet Pluto?

Explore the issues, and decide for yourself.

Arguments "FOR" Pluto being a planet:

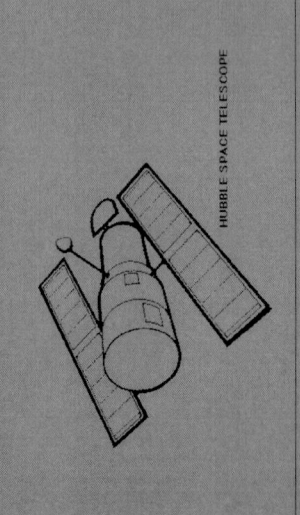

How many earth years does it take for Pluto to make one trip around the sun?

Explain the strange temperature on Pluto.

How long would it take a spacecraft to travel to Pluto?

What is the name of Pluto's moon?

What do some astronomers think this "moon" is?

Explain the mystery of whether Pluto is the 8th or 9th planet.

How has the Hubble Space Telescope helped in the study of Pluto?

HUBBLE SPACE TELESCOPE

Lesson 13

What is Polaris?

What is a star?

Brightest star?

Stars

Black Holes

Supernovas

Variable Stars

More Stars!

Lesson 13

Galaxy Shapes

Lesson 13

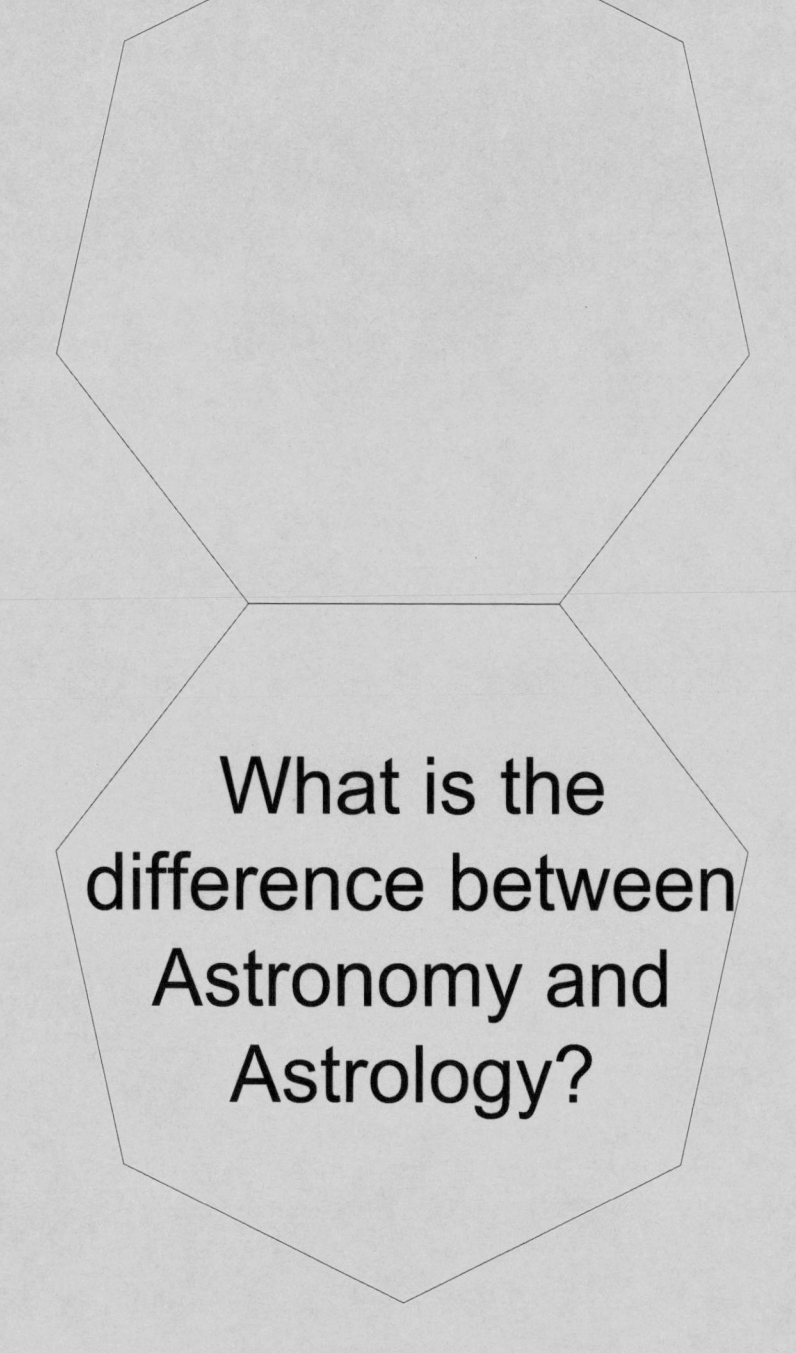

What is the difference between Astronomy and Astrology?

Lesson 13

How are Stars Categorized?

Hot or Cold

Suggested Paper Color: Yellow

Big or Small

Bright Or Dim

Suggested Paper Color: Yellow

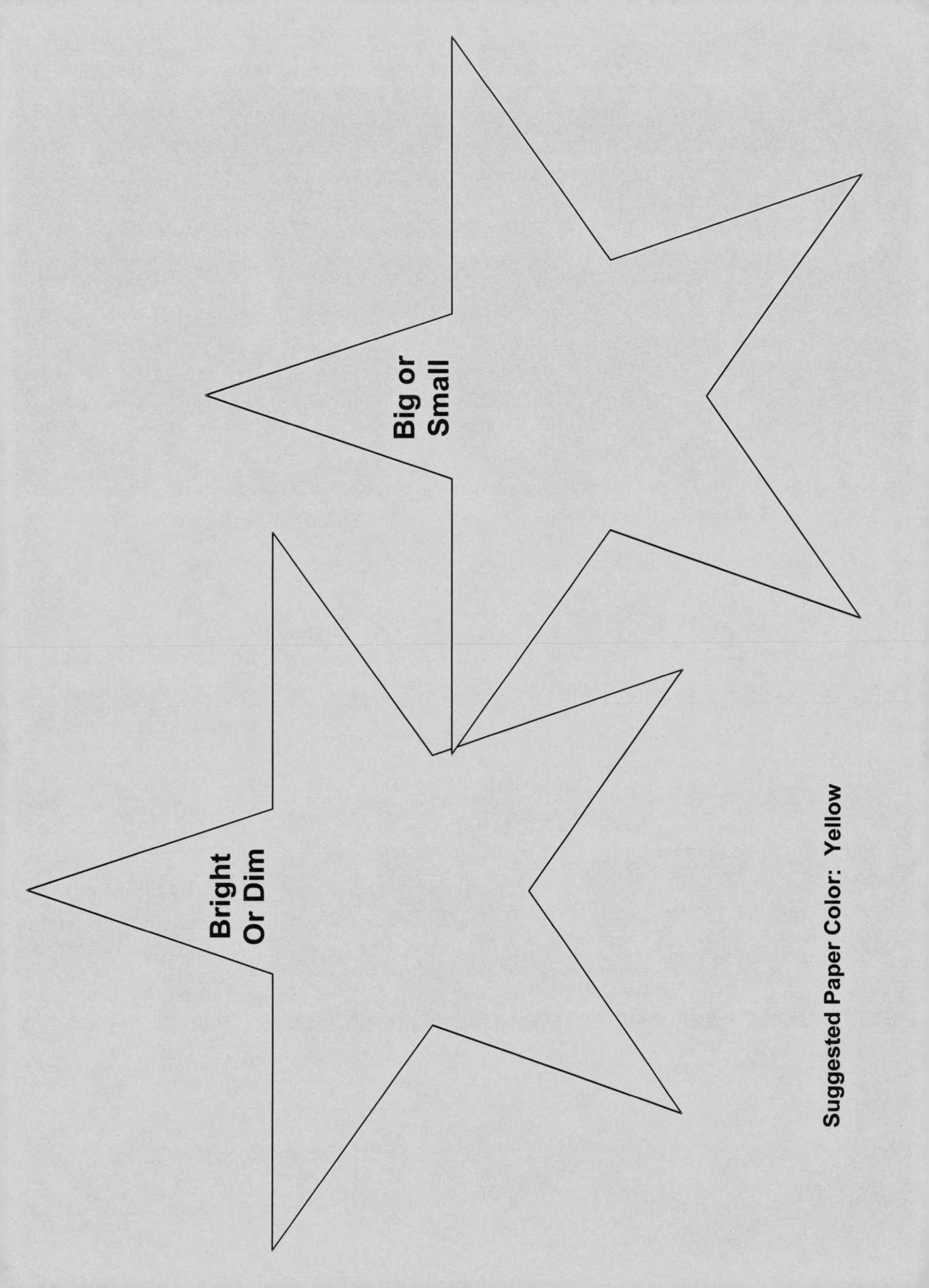

Suggested Paper Color: Yellow

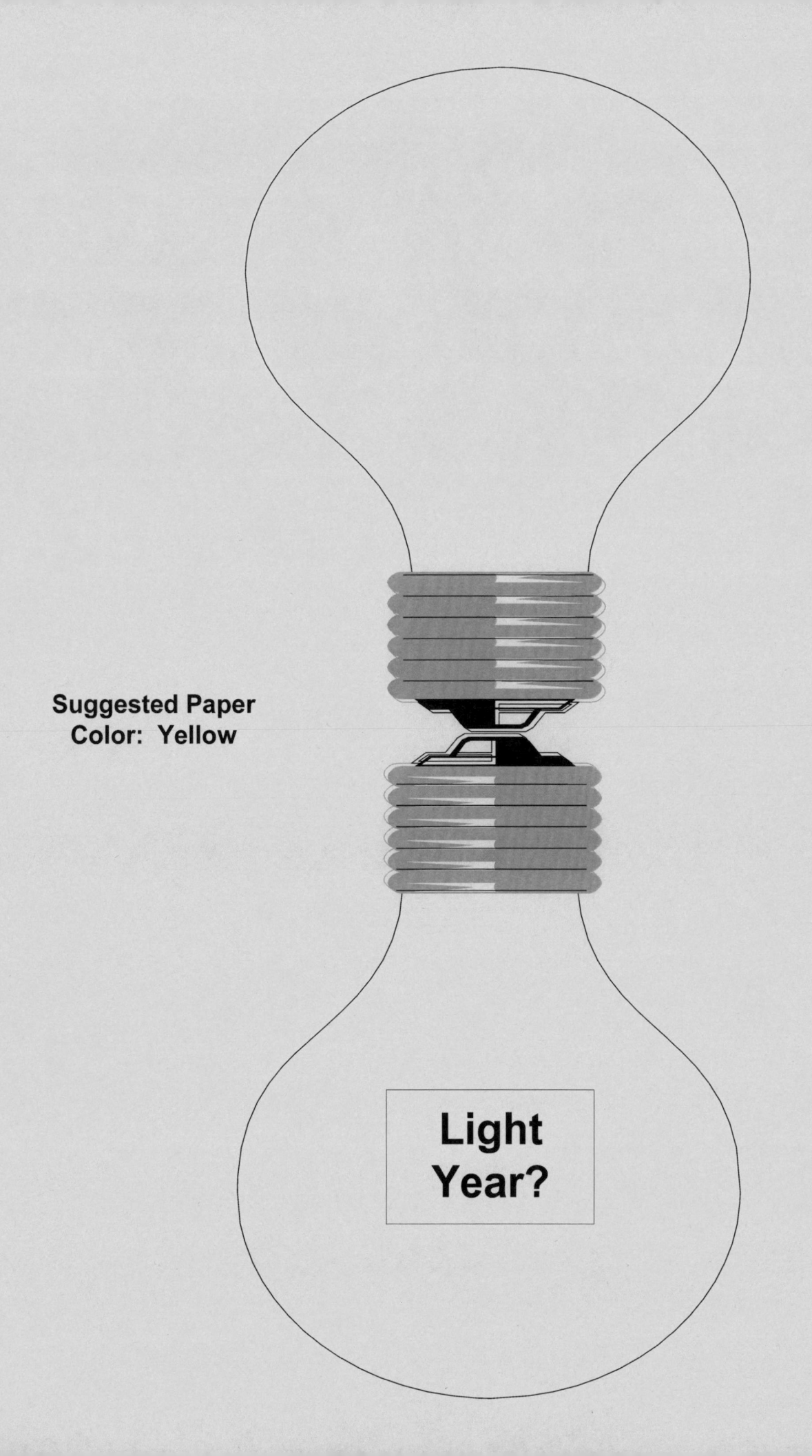

Light Year?

Sputnik & Space Race

Lesson 14

Lesson 14

Continue ON......... ↑

International Space Station

Lesson 14

Becoming a NASA Astronaut

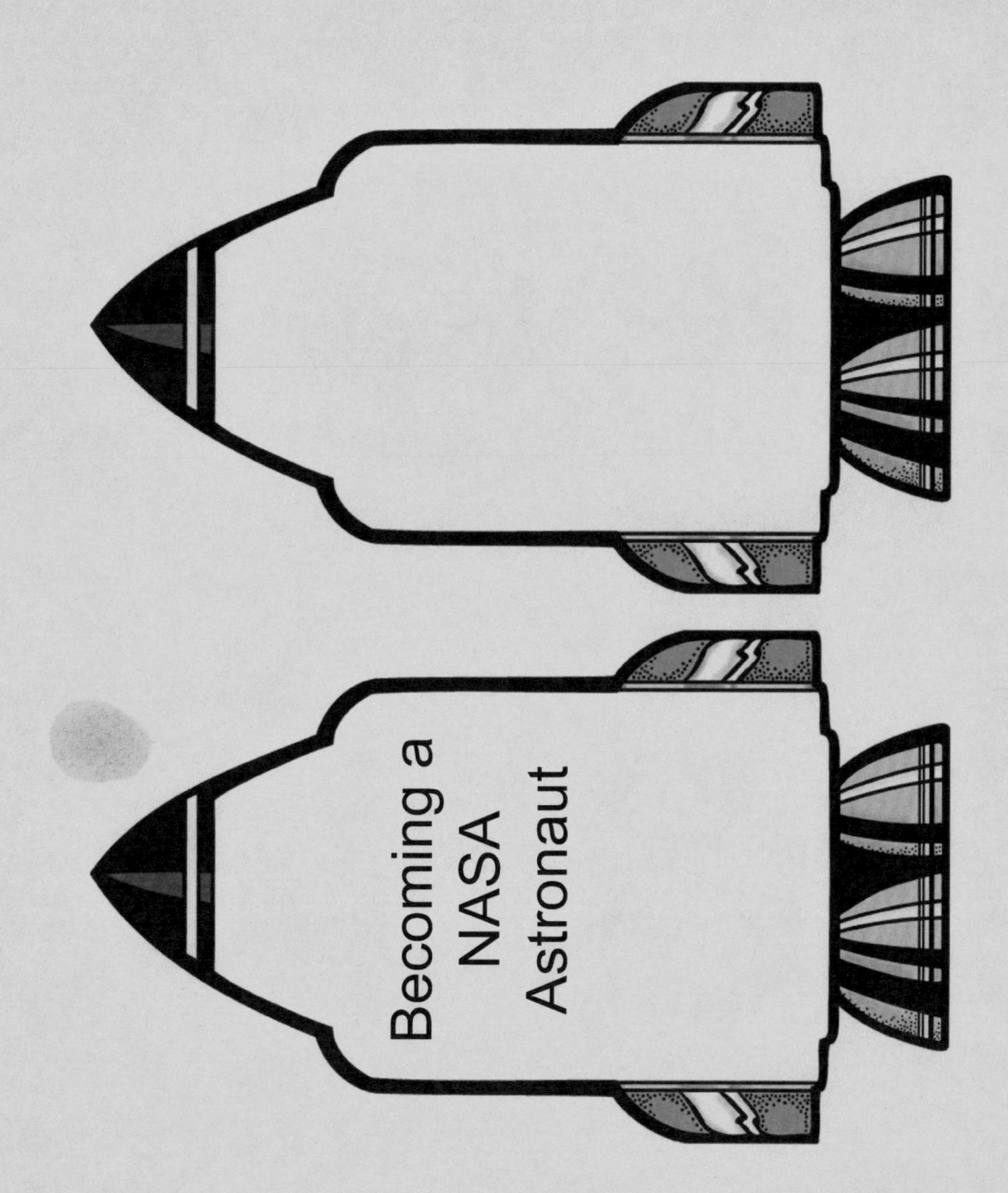

Lesson 14

If I could visit a planet, I would want to visit

Name the planets, in order (closest to the sun to farthest).

_____
_____
_____
_____
_____
_____
_____
_____
_____

Amazing Fact	Amazing Fact
Amazing Fact	Amazing Fact
Amazing Fact	